Understanding Volcanology

Understanding Volcanology

Ben Simmons

Callisto Reference

www.callistoreference.com

Callisto Reference,
118-35 Queens Blvd., Suite 400,
Forest Hills, NY 11375, USA

Visit us on the World Wide Web at:
www.callistoreference.com

ISBN: 978-1-64116-572-3 (Hardback)

Cataloging-in-Publication Data

Understanding volcanology / Ben Simmons.
p. cm.
Includes bibliographical references and index.
ISBN 978-1-64116-572-3
1. Volcanology. 2. Volcanoes. 3. Volcanism. 4. Geology. I. Simmons, Ben.
QE522 .U53 2022
551.21--dc23

Table of Contents

Preface

The purpose of this book is to help students understand the fundamental concepts of this discipline. It is designed to motivate students to learn and prosper. I am grateful for the support of my colleagues. I would also like to acknowledge the encouragement of my family.

The rupture in the crust of a planet such as Earth, which allows hot lava, gases and volcanic ash to escape from a magma chamber beneath the surface, is called a volcano. The discipline which is involved in the study of volcanoes, lava, magma and other related geophysical, geological and geochemical phenomena is called volcanology. It also studies the formation of volcanoes along with current as well as historic volcanic eruptions. There are various aspects related to volcanoes which are monitored under this field. A few of these are seismic observations, surface deformation monitoring and gas emission monitoring. Seismographs are deployed near volcanic areas to make seismic observations. Surface deformation monitoring observes the bulges near the volcanic center's surface which can indicate magma upwelling. Gas emissions are observed using equipment such as portable ultra-violet spectrometers to monitor changes in gas compositions, which can signal an impending volcanic eruption. This book provides comprehensive insights into the field of volcanology. The topics included herein on this subject are of utmost significance and bound to provide incredible insights to readers. Those in search of information to further their knowledge will be greatly assisted by this book.

A foreword for all the chapters is provided below:

Chapter – What is Volcano?

A volcano is described as a rupture in a planet's crust through which hot lava, volcanic ash, molten rock and gases get discharged from the magma chamber present below the surface. This chapter provides a brief introduction about volcanoes as well as the major areas of study related to them such as volcanic hotspots, volcanic ash and lava flows.

Chapter – Types of Volcano

Some of the major types of volcanoes are submarine volcanoes, strato volcanoes, shield volcanoes, cryovolcanoes, ice volcanoes, mud volcanoes, supervolcanoes and subglacial volcanoes. This chapter closely examines these types of volcanoes to provide an extensive understanding of the subject.

Chapter – Volcanology

The study of volcanoes, lava, magma and other geological phenomena related to volcanoes is known as volcanology. This field of geology analyzes all the major aspects of volcanoes such as volcanic arcs, volcanic belt, volcanic gas, lava balloons, etc. All these areas of study within volcanology have been carefully analyzed in this chapter.

Chapter - Volcanic Eruptions

A volcanic eruption is said to occur when hot materials like lava, molten rocks, gaseous compounds, dust, etc. are ejected from the Earth's crust. Some of the different types of volcanic eruptions are Hawaiian eruption, Strombolian eruption, Submarine eruption, Plinian eruption, etc. This chapter closely examines these types of volcanic eruptions as well as the causes of volcanic eruption.

Chapter - Plate Tectonics and Volcanism

The theory which describes the large-scale movement of the different plates in the Earth's lithosphere is termed as plate tectonics. Volcanism refers to the eruption of lava and volcanic gases through a rupture onto the Earth's surface. This chapter closely examines the key concepts of plate tectonics and volcanism.

Chapter - Impacts of Volcanoes

Volcanoes have a wide range of impacts upon the climate, animal life, plant life and soil. They are also responsible for shaping the surface of the Earth and generating new crust. The topics elaborated in this chapter will help in gaining a better perspective about these diverse impacts of volcanoes as well as their benefits.

Ben Simmons

1

What is Volcano?

A volcano is described as a rupture in a planet's crust through which hot lava, volcanic ash, molten rock and gases get discharged from the magma chamber present below the surface. This chapter provides a brief introduction about volcanoes as well as the major areas of study related to them such as volcanic hotspots, volcanic ash and lava flows.

Volcano is a vent in the crust of the Earth or another planet or satellite, from which issue eruptions of molten rock, hot rock fragments, and hot gases. A volcanic eruption is an awesome display of the Earth's power. Yet while eruptions are spectacular to watch, they can cause disastrous loss of life and property, especially in densely populated regions of the world. Sometimes beginning with an accumulation of gas-rich magma (molten underground rock) in reservoirs near the surface of the Earth, they can be preceded by emissions of steam and gas from small vents in the ground. Swarms of small earthquakes, which may be caused by a rising plug of dense, viscous magma oscillating against a sheath of more-permeable magma, may also signal volcanic eruptions, especially explosive ones. In some cases, magma rises in conduits to the surface as a thin and fluid lava, either flowing out continuously or shooting straight up in glowing fountains or curtains. In other cases, entrapped gases tear the magma into shreds and hurl viscous clots of lava into the air. In more violent eruptions, the magma conduit is cored out by an explosive blast, and solid fragments are ejected in a great cloud of ash-laden gas that rises tens of thousands of metres into the air. One feared phenomenon accompanying some explosive eruptions is the nuée ardente, or pyroclastic flow, a fluidized mixture of hot gas and incandescent particles that sweeps down a volcano's flanks, incinerating everything in its path. Great destruction also can result when ash collects on a high snowfield or glacier, melting large quantities of ice into a flood that can rush down a volcano's slopes as an unstoppable mudflow.

Strictly speaking, the term volcano means the vent from which magma and other substances erupt to the surface, but it can also refer to the landform created by the accumulation of solidified lava and volcanic debris near the vent. One can say, for example, that large lava flows erupt from Mauna Loa volcano in Hawaii, referring here to the vent; but one can also say that Mauna Loa is a gently sloping volcano of great size, the reference in this case being to the landform. Volcanic landforms have evolved over time as a result of repeated volcanic activity. Mauna Loa typifies a shield volcano, which is a huge, gently sloping landform built up of many eruptions of fluid lava. Mount Fuji in Japan is an entirely different formation. With its striking steep slopes built up of layers of ash and lava, Mount Fuji is a classic stratovolcano. Iceland provides fine examples of volcanic plateaus, while the seafloor around Iceland provides excellent examples of submarine volcanic structures.

Mount Fuji, Japan.

Volcanoes figure prominently in the mythology of many peoples who have learned to live with eruptions, but science was late in recognizing the important role of volcanism in the evolution of the Earth. Today geologists agree that volcanism is a profound process resulting from the thermal evolution of planetary bodies. Heat does not easily escape from large bodies such as the Earth by the processes of conduction or radiation. Instead, heat is transferred from the Earth's interior largely by convection—that is, the partial melting of the Earth's crust and mantle and the buoyant rise of magma to the surface. Volcanoes are the surface sign of this thermal process. Their roots reach deep inside the Earth, and their fruits are hurled high into the atmosphere.

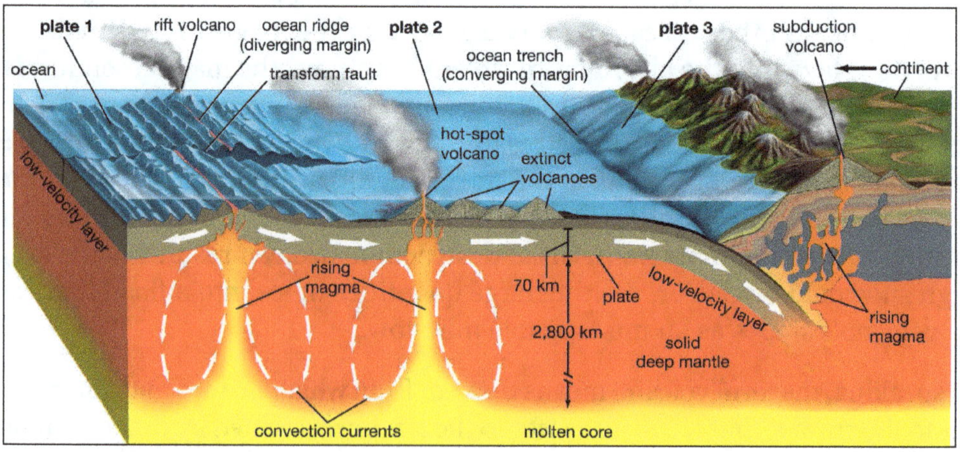

Volcanic activity and the Earth's tectonic plates stratovolcanoes tend to form at subduction zones, or convergent plate margins, where an oceanic plate slides beneath a continental plate and contributes to the rise of magma to the surface. At rift zones, or divergent margins, shield volcanoes tend to form as two oceanic plates pull slowly apart and magma effuses upward through the gap. Volcanoes are not generally found at strike-slip zones, where two plates slide laterally past each other. "Hot spot" volcanoes may form where plumes of lava rise from deep within the mantle to the Earth's crust far from any plate margins.

Volcanoes are closely associated with plate tectonic activity. Most volcanoes, such as those of Japan and Iceland, occur on the margins of the enormous solid rocky plates that make up the Earth's surface. Other volcanoes, such as those of the Hawaiian Islands, occur in the middle of a plate, providing important evidence as to the direction and rate of plate motion.

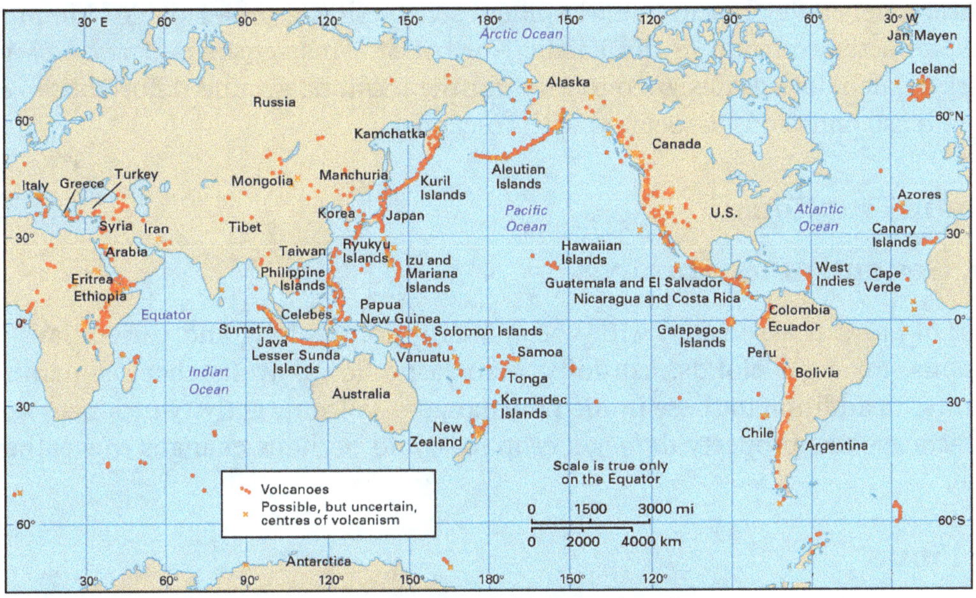

Volcanoes and thermal fields that have been active during the past 10,000 years.

The study of volcanoes and their products is known as volcanology, but these phenomena are not the realm of any single scientific discipline. Rather, they are studied by many scientists from several specialties: geophysicists and geochemists, who probe the deep roots of volcanoes and monitor signs of future eruptions; geologists, who decipher prehistoric volcanic activity and infer the likely nature of future eruptions; biologists, who learn how plants and animals colonize recently erupted volcanic rocks; and meteorologists, who determine the effects of volcanic dust and gases on the atmosphere, weather, and climate.

Clearly the destructive potential of volcanoes is tremendous. But the risk to people living nearby can be reduced significantly by assessing volcanic hazards, monitoring volcanic activity and forecasting eruptions, and instituting procedures for evacuating populations. In addition, volcanism affects humankind in beneficial ways. Volcanism provides beautiful scenery, fertile soils, valuable mineral deposits, and geothermal energy. Over geologic time, volcanoes recycle the Earth's hydrosphere and atmosphere.

In figure, a helicopter-borne "smart spider" sensor sitting on a ridge of Mount Saint Helens, an active volcano in the Pacific Northwest. This sensor is part of a wireless network of such devices designed to monitor the tremors, ground deformation, explosions, and ash emissions associated with volcanoes.

Volcanic Eruptions

Lava, Gas and other Hazards

The list of hazards associated with volcanic eruptions is long and varied: lava flows, explosions, toxic gas clouds, ash falls, pyroclastic flows, avalanches, tsunamis, and mudflows. In addition to these immediate dangers, volcanic activity produces secondary effects such as property damage, crop loss, and perhaps changes to weather and climate.

Lava Flows

The root zone of volcanoes is found some 70 to 200 km (40 to 120 miles) below the surface of the Earth. There, in the Earth's upper mantle, temperatures are high enough to melt rock and form magma. At these depths, magma is generally less dense than the solid rocks surrounding and overlying it, and so it rises toward the surface by the buoyant force of gravity. In some cases, as in the undersea zones where the tectonic plates of the Earth's crust are separating, magma may move directly up to the surface through fissures that reach as deep as the mantle. In other cases, it collects in large underground reservoirs known as magma chambers before erupting to the surface. Molten rock that reaches the surface is called lava.

Most magma formed by partial melting of the mantle is basaltic in composition, but, as it ascends, it assimilates silica, sodium, and potassium from the surrounding host rocks. Volcanic rocks found where magma erupts to the surface are classified into four major types, or "clans"—basalt, andesite, dacite, and rhyolite. These rocks are ranked, as can be seen in the table, mainly by their silica content, which ranges from approximately 50 percent for basalt to approximately 75 percent for rhyolite. As silica content increases, rock types generally become more viscous.

Table: Common Types of Volcanic Rock.

Name	Silica content (percent)	Major minerals	Colour	Approximate density with no voids (grams per cubic cm)
Obsidian glass can be dark gray to black.				
Basalt	45–53	Ca feldspar, pyroxene, olivine	dark gray	3.0
Andesite	53–62	CaNa feldspar, pyroxene, amphibole	medium gray	2.9
Dacite	62–70	Na feldspar, amphibole, biotite, quartz	light gray to tan	2.8
Rhyolite	70–78	K, Na feldspars, quartz, biotite	light gray to pink	2.7

If the vast, unseen undersea lava flows of the oceanic ridge system are considered, lava flows are the most common products of the Earth's volcanoes. There are two major types of lava flow, referred to around the world by their Hawaiian names: pahoehoe, a more fluid flow with a smooth to ropy surface; and aa (or a'a), a more viscous flow whose surface is covered by thick, jumbled piles of loose, sharp blocks. Both types have the same chemical composition; the difference seems to be in the eruptive temperature and the speed of movement of the flow. As much as 99 percent of the island of Hawaii is composed of aa and pahoehoe flows. Indeed, Kilauea volcano has erupted continuously since 1983, its lava flows covering more than 100 square km (40 square miles) of land and adding more than 2 square km (0.8 square mile) to the island where the lava has poured into the ocean. In the Mediterranean region, Mount Etna has issued lava more than 150 times since its first recorded activity in 1500 BC.

Pahoehoe lava flow, Kilauea volcano, Hawaii, November.

Explosions

Massive volcanic explosions are caused by the rapid expansion of gases, which in turn can be triggered by the sudden depressurization of a shallow hydrothermal system or gas-charged magma body or by the rapid mixing of magma with groundwater. The ash, cinders, hot fragments, and bombs thrown out in these explosions are the major products observed in volcanic eruptions around the world. These solid products are classified by size. Volcanic dust is the finest, usually about the consistency of flour. Volcanic ash is also fine but more gritty, with particles up to the size of grains of rice. Cinders, sometimes called scoriae, are the next in size; these coarse fragments can range from 2 mm (0.08 inch) up to about 64 mm (2.5 inches). Fragments larger than 64 mm are called either blocks or bombs. Volcanic blocks are usually older rock broken by the explosive opening of a new vent. Large blocks ejected in such explosions have been hurled as far as 20 km (12 miles) from the vent. Volcanic bombs, in contrast, are generally incandescent and soft during their flight. Some bombs take on strange, twisted shapes as they spin through the air. Others have a cracked and separated crust that has cooled and hardened in flight; they are called "breadcrust bombs."

A cloud of ash and pumice rises into the air on July 22, 1980, following an explosive eruption of Mount St. Helens, Washington state, U.S.

A directed blast in which one side of a volcanic cone fails, as happened at Mount St. Helens in the United States in 1980, can cause destruction over several hundred square kilometres on the failed flank of the volcano. This is especially true if the blast cloud is heavily laden with fragmental debris and becomes dense and fluidized. It then takes on characteristics similar to a pyroclastic flow.

Pyroclastic Flows

Pyroclastic flows are the most dangerous and destructive aspect of explosive volcanism. Variously called nuées ardentes ("glowing clouds"), glowing avalanches, or ash flows, they occur in many sizes and types, but their common characteristic is a fluidized emulsion of volcanic particles, eruption gases, and entrapped air, resulting in a flow of sufficiently low viscosity to be very mobile and of sufficiently high density to hug the ground surface. A pyroclastic flow can pour over the lip of an erupting vent, or it may form when an ash column becomes too dense to continue rising and falls back to the ground. In major caldera collapses associated with explosive volcanoes, huge pyroclastic flows may issue from the ring fractures as the caldera block subsides.

Pyroclastic flows can move at speeds up to 160 km (100 miles) per hour and have temperatures ranging from 100 to 700 °C (212 to 1,300 °F). They sweep away and incinerate nearly everything in their path. Smaller pyroclastic flows are often confined to valleys. Large pyroclastic flows may spread out as a blanket deposit across many hundreds or even thousands of square kilometres around a major caldera collapse. During the past two million years, the area around Yellowstone National Park in the western United States has undergone three major caldera collapses involving pyroclastic eruptions of 280 to 2,500 cubic km (67 to 600 cubic miles) of ash flows and ash falls.

Gas Clouds

Even beyond the limit of explosive destruction, the hot, ash-laden gas clouds associated with an explosive eruption can scorch vegetation and kill animals and people by suffocation. Gas clouds emitted from fumaroles (volcanic gas vents) or from the sudden overturn of a crater lake may contain suffocating or poisonous gases such as carbon dioxide, carbon monoxide, hydrogen sulfide, and sulfur dioxide. At Lake Nyos, a crater lake in Cameroon, West Africa, more than 1,700 people were killed by a sudden release of carbon dioxide in August 1986. Scientists theorize that carbon dioxide of volcanic origin had been seeping into the lake, perhaps for centuries, and had accumulated in its deep layers. It is thought that some disturbance, such as a large landslide into the lake, could have triggered the outburst of gas, creating an effervescence that stirred the lake and started the degassing.

The most common volcanic gases are water vapour, carbon dioxide, sulfur dioxide, and hydrogen sulfide. Small quantities of other volatile elements and compounds also are present, such as hydrogen, helium, nitrogen, hydrogen chloride, hydrogen fluoride, and mercury. The specific gaseous compounds released from magma depend on the temperature, pressure, and overall composition of the volatile elements present. The amount of available oxygen is of critical importance in determining which volatile gases are present. When oxygen is lacking, methane, hydrogen, and hydrogen sulfide are chemically stable, but when hot volcanic gases mix with atmospheric gases, water vapour, carbon dioxide, and sulfur dioxide are stable.

Some volcanic gases are less soluble in magma than others and therefore separate at higher pressures. Studies at Kilauea in Hawaii indicate that carbon dioxide begins to separate from its parent magma at depths of about 40 km (25 miles), whereas most of the sulfur gases and water are not released until the magma has nearly reached the surface. Fumaroles near Halemaumau Crater at Kilauea's summit are rich in carbon dioxide that leaks from the magma chamber located 3 to 4 km (1.9 to 2.5 miles) beneath the surface. Fumaroles on the rift zones of Kilauea, however, are richer in water vapour and sulfur because much of the carbon dioxide leaks away at the summit before the magma is intruded into the rift zones.

Ash Falls

Ash falls from continued explosive jetting of fine volcanic particles into high ash clouds generally do not cause any direct fatalities. However, where the ash accumulates more than a few centimetres, collapsing roofs and failure of crops are major secondary hazards. Crop failure can occur over large areas downwind from major ash eruptions, and widespread famine and disease may result, especially in poorly developed countries. In the long run, however, the decomposition of nutrient-rich volcanic fallout is responsible for some of the world's best soils.

Buildings and vegetation at Clark Air Base, Philippines, destroyed by a thick, wet layer of ash following the gigantic explosion of Mount Pinatubo.

Avalanches, Tsunamis and Mudflows

Avalanches of rock and ice also are common on active volcanoes. They may occur with or without an eruption. Those without an eruption are often triggered by earthquakes, by weakening of rock into clay by hydrothermal activity, or by heavy rainfall or snowfall. Those associated with eruptions are sometimes caused by oversteepening of a volcano's flank by intrusion of a shallow body of magma within or just beneath the volcanic cone, as happened at Mount St. Helens.

A caldera collapse that is in part or entirely submarine usually generates a tsunami. The larger and more rapid the collapse, the larger the tsunami. Tsunamis also can be caused by avalanches or large pyroclastic flows rapidly entering the sea on the flank of a volcano.

Mudflows, or lahars, are common hazards associated with stratovolcanoes and can happen even without an eruption. They occur whenever floods of water mixed with ash, loose soil, or hydrothermal clay sweep down valleys that drain the sides of large stratovolcanoes. The huge mudflows generated by meltwater from the ice cap of Mount Ruiz, Colombia, in 1985 are classic examples of mudflows associated with eruptions. Heavy rainfall or earthquake-induced avalanches of ice or hydrothermal clay also can cause mudflows on steep volcanoes during periods of repose between eruptions.

Secondary Damage

Property damage from volcanic eruptions is difficult to estimate, because of differing value systems and changes in land use. One study estimates an average of $1 billion per year in property damage worldwide from volcanic eruptions. As with casualties, a few eruptions cause staggering damage, while most are much less destructive. The Mount St. Helens eruption in 1980 caused more than $1 billion worth of damage, mainly to the timber industry. The economic cost of the 1991 eruption of Mount Pinatubo in the Philippines was estimated at $7 billion, though it is likely that losses continued to mount for years afterward because of the inundation of much arable land by mudflows.

A new danger that emerged with the increase in air travel is the serious threat posed to jet aircraft by high clouds of volcanic ash and aerosols. These clouds cannot be detected by weather radar, and they are difficult for pilots to distinguish from meteorological clouds. In a small number of documented cases, jet engines have stopped operating after airliners have flown through volcanic clouds. Catastrophe was avoided only at the last minute when the pilots were able to restart the engines as their planes descended below the clouds. The engines had to be replaced and major abrasive damage to the planes repaired. During the Pinatubo eruption, aircraft had 16 damaging encounters with ash clouds, one as far away as 1,700 km (1,050 miles) west of the volcano.

In response to these hazards, a worldwide system was established to alert pilots by radio about volcanic eruption clouds and their probable extent. This system, which is made up of a network of nine Volcanic Ash Advisory Centres (VAACs) managed by the International Civil Aviation Organization, helps aviation officials divert air traffic around areas of dangerous ash concentrations. A few weeks after the start of the eruption of Iceland's Eyjafjallajökull volcano in March 2010, ash plume data gathered from the London VAAC factored into the decision by national aviation authorities to ground flights for several days throughout northern and central Europe.

Long-term Environmental Effects

Not all volcanic phenomena are destructive. The oceans, atmosphere, and continents owe their origin and evolution in large measure to volcanic processes throughout geologic time. A lava flow may engulf and bury the land, but new soil and vegetation eventually develop. In warm, humid climates the recovery is rapid; a few decades will suffice to hide the rocky surface of solidified lava flows. In desert or Arctic climates, on the other hand, recovery is slower; flows more than 1,000 years old may still retain their barren appearance. Volcanic ash slowly weathers to form rich, loamy soils. On the volcanic island of Java, terraced rice paddies support a dense population. Across the Java Sea is Borneo, an island with a similar climate but no volcanoes. The jungles of Borneo provide only temporary slash-and-burn agriculture and support a much smaller population.

Climate, too, is subject to the effects of volcanic activity. High ash clouds, especially if they are rich in sulfur dioxide, can inject much fine dust and aerosol droplets of sulfuric acid into the stratosphere, above tropospheric rain clouds. Their height greatly increases the residence time of these fine particles in the atmosphere—they are not washed quickly back to Earth but spread slowly into haze layers that can blanket a hemisphere or even the entire Earth.

Pinatubo, Mount: Earth's atmosphere showing the Mount Pinatubo dust layer, photographed from the U.S. space shuttle Atlantis.

World climate seems to have been affected by the eruptions of Krakatoa (Krakatau) near Java in 1883, Mount Agung in Bali in 1963, and Pinatubo in 1991. The high ash clouds thrown up by these volcanoes apparently lowered average world temperature by about 0.5 °C (0.9 °F) over one to three years following their eruptions. Although world temperature data was poorly recorded in the early 1800s, the eruption of Mount Tambora on the island of Sumbawa in 1815 was followed in 1816 in North America and Europe by what was called "the year without a summer." On the other hand, other large eruptions, such as Novarupta near Mount Katmai in Alaska in 1912, appear to have produced no cooling effect. Records of average world temperature over the past several decades often show changes of 0.1 to 0.3 °C (0.2 to 0.5 °F) from year to year unrelated to any known volcanic eruptions, so it is difficult to establish with certainty whether volcanoes have a major impact on climate.

Direct sampling of the stratosphere has shown that the major haze-forming agent from volcanic eruptions is not fine dust but an aerosol of tiny sulfuric acid droplets. This indicates that the composition of high volcanic ash clouds may be as important as their volume in affecting climate. Atmospheric chemists are interested in atmospheric perturbations that may be caused not only by volcanic eruptions but also by man-made aerosols of chlorofluorocarbons, exhaust from high-altitude jet aircraft, and a general increase in carbon dioxide and other greenhouse gases from the burning of fossil fuels. The Earth has many buffers that maintain its environment, but their interactions are not clearly understood. Many questions as to how volcanic and human activity affect climate remain largely unanswered, and they are important problems of ongoing research.

Six Types of Eruptions

Volcanoes are frequently classified by their size and shape, but they can also be classified by their eruptive habits. Indeed, the type of volcanic eruption that occurs plays an important role in the evolution of a volcanic landform, thus forming a significant link between eruptive habit and volcanic structure. In general, eruptions can be categorized as either effusive or explosive. Effusive eruptions involve the outpouring of basaltic magma that is relatively low in viscosity and in gas content. Explosive eruptions generally involve magma that is more viscous and has a higher gas content. Such magma is often shattered into pyroclastic fragments by explosive gas expansion during an eruption.

In more-detailed classification schemes based on character of eruption, volcanic activity and volcanic areas are commonly divided into six major types, shown schematically in the diagram. They are listed as follows in order of increasing degree of explosiveness:

1. Icelandic,
2. Hawaiian,
3. Strombolian,
4. Vulcanian,
5. Pelean,
6. Plinian.

The Icelandic type is characterized by effusions of molten basaltic lava that flow from long, parallel fissures. Such outpourings often build lava plateaus.

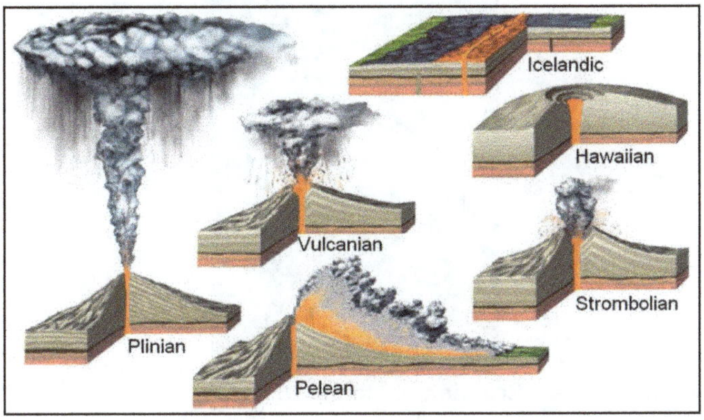

The major types of volcanic eruptions.

The Hawaiian type is similar to the Icelandic variety. In this case, however, fluid lava flows from a volcano's summit and radial fissures to form shield volcanoes, which are quite large and have gentle slopes.

Strombolian eruptions involve moderate bursts of expanding gases that eject clots of incandescent lava in cyclical or nearly continuous small eruptions. Because of such small frequent outbursts, Stromboli volcano, located on Stromboli Island off the northeast coast of Italy, has been called the "lighthouse of the Mediterranean."

The Vulcanian type, named for Vulcano Island near Stromboli, generally involves moderate explosions of gas laden with volcanic ash. This mixture forms dark, turbulent eruption clouds that rapidly ascend and expand in convoluted shapes.

A Pelean eruption is associated with explosive outbursts that generate pyroclastic flows, dense mixtures of hot volcanic fragments and gas. Pelean eruptions are named for the destructive eruption of Mount Pelée on the Caribbean island of Martinique in 1902. The fluidized slurries produced by these eruptions are heavier than air but are of low viscosity and pour down valleys and slopes at great velocities. As a result, they are extremely destructive.

The Plinian type is an intensely violent kind of volcanic eruption exemplified by the outburst of Mount Vesuvius in Italy in AD 79 that killed the famous Roman scholar Pliny the Elder and was described in an eyewitness account by his nephew, the historian Pliny the Younger. In this type of eruption, gases boiling out of gas-rich magma generate enormous and nearly continuous jetting blasts that core out the magma conduit and rip it apart. The uprushing gases and volcanic fragments resemble a gigantic rocket blast directed vertically upward. Plinian eruption clouds can rise into the stratosphere and are sometimes continuously produced for several hours. Lightning strikes caused by a buildup of static electricity are common close to Plinian ash clouds, adding one more element of terror to the eruption.

Eruption of Mount St. Helens.

Why are some volcanic eruptions so explosive while others are so spectacular but relatively harmless? The answer involves at least four factors: the amount of gas dissolved in the magma, the viscosity of the magma, the rate of decompression of the magma as it rises toward the surface, and the number of nucleation sites on which the gases can begin to form bubbles. Volcanoes related to converging plate margins generally have a high gas content, and their magma is very viscous. This combination is explosive because the gases cannot easily boil out; rather, they remain pent up until they reach the pressure at which they blow the viscous magma into fragments. The rate at which pressure is reduced also controls the explosiveness. If magma moves slowly toward the surface, its dissolved gases will be released slowly and can escape. During the 1991 Plinian-type eruption of Mount Pinatubo, magma moved quite rapidly toward the surface, resulting in retention of most of the dissolved gases. Finally, the speed at which gases are released from magma is affected by the number of small crystals, which can act as nucleation sites where gas bubbles begin to form. At Pinatubo the magma was more than 40 percent small crystals before the eruption, while at the Hawaiian volcanoes Kilauea and Mauna Loa the percentage of small crystals in the magma is very low (less than 5 percent).

Volcano Forecasting and Warning

The greatest hazard at potentially active volcanoes is human complacency. The physical hazards can be reliably estimated by studying past eruptive activity as recorded in history or in the prehistoric deposits around a volcano. Volcano observatories can monitor local earthquake activity and the surface deformation of a potentially active volcano and make useful, if not yet precise, forecasts of eruptions. For instance, the measurement of increased earthquake activity beneath Mauna Loa in 1983 led to a forecast of an increase in probability of an eruption for 1984 or 1985; an eruption occurred in March 1984. The major eruption of Mount St. Helens on May 18, 1980, was much larger than anticipated, but a high number of local earthquakes and a visible bulge forming on the north flank of the mountain provided enough warning to encourage a partial evacuation of the surrounding area. Lives were lost, but the toll would have been much higher if access to the area had not been restricted by local authorities. A major problem in reducing volcanic risk is that most explosive volcanoes have such long repose periods that people living nearby consider them extinct rather than dormant.

The correct forecast and evacuation of residents before the 1991 eruption of Mount Pinatubo saved thousands of lives, but the science of eruption forecasting is only beginning and is still fraught with uncertainty. Evacuation of large numbers of people is difficult and expensive. A major evacuation not followed by any major eruption would be a serious mistake, but not evacuating people from a threatening volcano that then erupts catastrophically would be a much worse mistake. It is not a simple problem.

Volcanic Landforms

Pyroclastic Cones

Pyroclastic cones (also called cinder cones or scoria cones) such as Cerro Negro in Nicaragua are relatively small, steep (about 30°) volcanic landforms built of loose pyroclastic fragments, most of which are cinder-sized. The fragments cool sufficiently during their flight through the air so that they do not weld together when they strike one another. Generally, the crater from which the cinder fragments were ejected is located in the centre of the cone. In areas with strong prevailing winds, however, the crater may be upwind of the cone. The rock type involved in pyroclastic cones is generally basalt or basaltic andesite, and the eruption type is either the moderately explosive Vulcanian or the gentler Hawaiian, which produces high lava fountains.

Cerro Negro is a cinder cone type of volcano that was born of a series of eruptions beginning in 1850. It still periodically blankets the surrounding countryside with ash.

Some cinder cones such as Parícutin in Mexico grow during a single eruption. Parícutin rises approximately 410 metres (1,345 feet) from its base to its summit and is 1 km (0.6 mile) wide; it formed during nearly continuous eruptions from 1943 to 1952. Cinder cones also form at some vents on shield volcanoes, but these are not considered to be separate, individual volcanoes. Certain cinder cones have multiple eruptions, but, if activity continues for thousands to tens of thousands of years from the same vent, it is likely that they will develop into stratovolcanoes or complex volcanoes.

Eruption of Cerro Negro volcano, Nicaragua.

Pumice cones are structures similar to cinder cones, but they are made up of volcanic glass fragments so riddled with gas-bubble holes (vesicles) that they resemble a sponge and are very lightweight. Less common pyroclastic landforms include maars, low-relief craters often filled with water and surrounded by a rim of ejected material that was probably formed by explosive interaction of magma and groundwater; and tuff rings and tuff cones, which are landforms built of compacted pyroclastic deposits. Tuff rings and cones resemble maars, but they have higher rims and are not filled with water. Tuff rings are only about 5 metres (16 feet) high, with craters roughly at ground level. Tuff cones are higher and steeper, with craters above ground level. Punchbowl and Diamond Head on Oahu island, Hawaii, are famous examples of tuff cones.

Volcanic Fields

Such areas have many geologically young cinder cones or other features that have not been individually identified as separate volcanoes. If the conduits through which magma ascends to the surface are scattered over a broad area, many short-lived volcanoes are formed rather than a major volcano with repeated eruptions. The area in which Parícutin formed is a volcanic field with dozens of prehistoric—but geologically young—cinder cones and lava flows. The most likely place for the birth of a new volcano is in a known volcanic field.

Fissure Vents

These features constitute the surface trace of dikes (underground fractures filled with magma). Most dikes measure about 0.5 to 2 metres (1.5 to 6.5 feet) in width and several kilometres in length. The dikes that feed fissure vents reach the surface from depths of a few kilometres. Fissure vents are common in Iceland and along the radial rift zones of shield volcanoes.

In Iceland the volcanic vents often are long fissures parallel to the rift zone where lithospheric plates are diverging. Renewed eruptions generally occur from new parallel fractures offset by a few hundred to thousands of metres from the earlier fissures. This distribution of vents and voluminous eruptions of fluid basaltic lava usually build up a thick lava plateau rather than a single volcanic edifice. The largest effusive eruption of lava in recorded history occurred in 1783 in Iceland from the Laki fissure. This vent produced high lava fountains, a crater row 25 km (15.5 miles) long, and 565 square km (218 square miles) of basaltic lava flows with a volume of approximately 12 cubic km (2.9 cubic miles).

The radial fissure vents of Hawaiian volcanoes produce "curtains of fire" as lava fountains erupt along a portion of a fissure. These vents produce low ramparts of basaltic spatter on both sides of the fissure. More isolated lava fountains along the fissure produce crater rows of small spatter and cinder cones. The fragments that form a spatter cone are hot and plastic enough to weld together, while the fragments that form a cinder cone remain separate because of their lower temperature.

Lava Domes

Landforms of this sort consist of steep domal mounds of lava so viscous that the lava piles up over its vent without flowing away. The rock types that form lava domes are generally andesites, dacites, or rhyolites. Somehow these viscous lavas have lost much of their gas content in prior eruptions or during a slow rise to the surface. Even so, it is not unusual for an actively growing lava dome to have an explosive eruption that disrupts all or part of the dome. Many lava domes grow by internal intrusion of lava that causes swelling and oversteepening of the dome. Rockslides build up an apron of talus blocks around the lower sides of the dome. Lava domes can form mounds several hundred metres high with diameters ranging from several hundred to more than 1,000 metres (3,300 feet). Thick lava flows sometimes move short distances from the dome and distort its generally circular or oval shape. A good example of a lava dome is the one in the explosion crater at Mount St. Helens.

The lava dome of Mount St. Helens.

Following the great eruption of May 1980, a dome of lava grew intermittently in the crater of the volcano. By the time of this photograph, the dome measured 850 metres (2,800 feet) wide and 220 metres (725 feet) high.

Other Volcanic Structures and Features

There are many types of volcanic forms and terms other than those described above. Some general terms that may be encountered include volcanic cone, which is a descriptive term pertaining to shape with no implication of size, rock type, or genesis; and explosion crater, a large circular, elongate, or horseshoe-shaped excavation with ejected debris on its rim or flanks. A somma volcano, named for Mount Somma, a ridge on the slopes of Mount Vesuvius in Italy, is a caldera partially filled by a new central cone. In some areas, magma or still-hot igneous rocks at shallow depth leak gases through gas vents or interact with the groundwater system to create hot springs. These areas are known as hydrothermal regions, fumarole fields, or solfatara fields.

Determinants of Size and Shape

The shape and size of a volcano are controlled by several factors. These include:

1. The volume of volcanic products.
2. The interval length between eruptions.
3. The composition of volcanic products.
4. The variety of volcanic eruption types.
5. The geometry of the vent.
6. The environment into which the volcanic products are erupted.

The volume of material released in any one eruption can vary enormously from a few cubic metres of magma to as much as 3,000 cubic km (720 cubic miles). A series of small eruptions usually builds up mounds close to the vent, whereas large-volume eruptions tend to disperse their products over a greater distance. Effusive eruptions form lava plateaus or gently sloping shield volcanoes; moderately explosive eruptions form stratovolcanoes; and giant explosive eruptions form plateaus of lava or ash flows and almost always form a caldera several kilometres in diameter over the eruption site. Naturally, since many other factors are involved in determining volcanic landforms, there are exceptions to these rules.

The chemical composition of magma affects its physical properties, which in turn have a major influence on the landform built by a volcanic eruption. Four common volcanic rock types are listed in the table: basalt, andesite, dacite, and rhyolite. As the silica content increases, these rock types generally become more viscous; as the magmatic gas content increases, they become more explosive. Other physical properties are, however, important in determining the character of lava flows. For example, hot basaltic lava produces flows with smooth to ropy surfaces. These flows, known as pahoehoe, tend to flow farther than the cooler aa flows of the same chemical composition that have rough, broken surfaces.

If a volcano has consistent eruption habits, its landform will reflect that character. The shape of the huge but gently sloping shield volcano Mauna Loa, for example, indicates a long record of eruption of fluid lava flows, while the beautiful, symmetrical shape of the stratovolcano Mount Fuji indicates a long record of moderately explosive eruptions from its summit that produce alternating layers of ash and lava. In contrast to simple shield and stratovolcanoes, many volcanoes change their eruptive habits—both in eruption type and in the location of their vents—over time. This results in a mixture of volcanic landforms called a complex volcano.

The geometry of the vent or vents also exerts profound control on volcanic landforms. Multiple point-source vents that erupt only once or at most a few times form a volcanic field of dozens of small cinder cones rather than a single large volcanic edifice. Volcanic vents in the form of long fissures usually build up a thick lava plateau (especially when erupting voluminous amounts of fluid lava) or a low volcanic mound. Hekla Volcano in Iceland is transitional between plateau-building fissure eruptions and eruptions from a single major vent that produce a symmetrical stratovolcano. Hekla erupts from a fissure that is parallel to the Mid-Atlantic Ridge and about 20 km (12 miles) in length. Viewed along the fissure, Hekla looks like a stratovolcano; perpendicular to the fissure, it appears as an elongate ridge.

Finally, but of great importance, is the environment where the volcanic products are erupted—into the atmosphere, under water, or under ice. Submarine volcanoes are surprisingly similar to their counterparts on land, but their slopes are generally steeper because water cools the lavas more rapidly. Deep submarine volcanism tends to be less explosive because the pressure of the water retards explosive boiling. Subglacial volcanism produces landforms that are dramatically different from those produced by subaerial volcanism. This is particularly apparent in Iceland, where glaciers covered the entire island 15,000 years ago, and large ice caps still cover extensive areas today. Fissure eruptions beneath the ice form steep ridges of broken lava fragments rather than lava-flow plateaus, while subglacial eruptions from point-source vents that erupt repeatedly form table mountains. Table mountain volcanoes have steep sides of pillow lavas—sacklike structures that form when flows of basaltic lava are extruded into the ocean, a deep lake, or a water-filled cavern within ice. These pillow structures are capped by several tens of metres of broken lava fragments from explosive shallow-water eruptions. The broken lava fragments in turn are overlain by shield-building lava flows erupted above the glacial surface.

Hot Springs and Geysers

Hot springs and geysers also are manifestations of volcanic activity. They result from the interaction of groundwater with magma or with solidified but still-hot igneous rocks at shallow depths.

Yellowstone National Park in the United States is one of the most famous areas of hot springs and geysers in the world. The total heat flux from these thermal features is estimated to be 300 megawatts (300 million watts). The last great eruption at Yellowstone occurred about 630,000 years ago when some 1,000 cubic km (240 cubic miles) of rhyolitic pumice and ash were ejected in huge pyroclastic flows and resulted in the formation of a caldera—a large circular or oval depression caused by collapse of the surface following magma removal—approximately 45 by 75 km (28 by 47 miles) in size. Yellowstone Lake now occupies part of this giant caldera. Since that last great outburst, about 1,200 cubic km (288 cubic miles) of rhyolite lava flows and domes have erupted in numerous smaller events. The cooling roots of such past eruptions, or possibly the new intrusions of magma at shallow depth, are the heat sources for the Yellowstone hot springs and geysers.

Geysers are hot springs that intermittently spout a column of hot water and steam into the air. This action is caused by the water in deep conduits beneath a geyser approaching or reaching the boiling point. At 300 metres (about 1,000 feet) below the surface, the boiling point of water increases to approximately 230 °C (450 °F) because of the increased pressure of the overlying water. As bubbles of steam or dissolved gas begin to form, rise, and expand, hot water spills from the geyser's vent, lowering the pressure on the water column below. Water at depth then momentarily exceeds its boiling point and flashes into steam, forcing additional water from the vent. This chain reaction continues until the geyser exhausts its supply of boiling water.

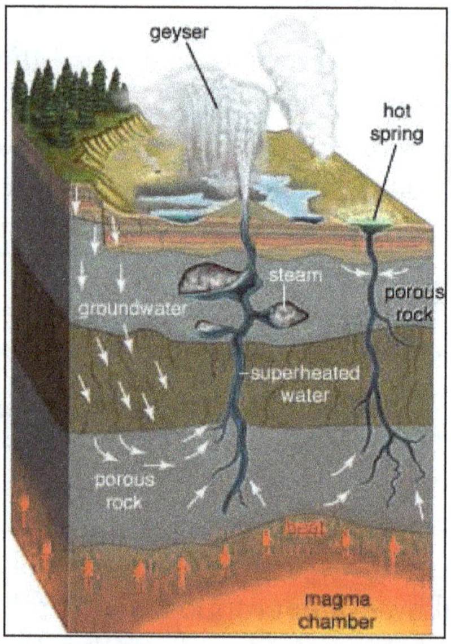

Cross section of a geyser and hot spring.

Groundwater percolates through porous rock into fractures deep underground, where heat from a nearby magma chamber superheats the pressurized water to a temperature above the boiling point of water at surface pressure. In hot springs the rising superheated water is cooled below the boiling point by groundwater before reaching the surface. In geysers the superheated water collects in underground pockets. There a small drop in pressure caused by the release of water at the surface flashes the superheated water into steam, which expands and ejects a column of steam and water into the air. When the supply of steam and hot water is exhausted, the spouting stops and the cycle begins again.

Old Faithful geyser, Upper Geyser Basin, Yellowstone National Park, Wyoming, U.S.

After a geyser stops spouting, the conduits at depth refill with groundwater, and reheating begins again. In geysers such as Yellowstone's Old Faithful, the spouting and recharge period is quite regular. This famous geyser has gushed to heights of 30 to 55 metres (100 to 180 feet) about every 90 minutes for more than 100 years. If Old Faithful's eruption lasts only a minute or two, the next interval will be shorter than average, while a four-minute eruption will be followed by a longer interval. Other geysers have much more erratic recharge times.

Volcanism and Tectonic Activity

Taking a more distant view of volcanic landforms from space, one can see that most volcanoes group together to form linear to arcuate belts across the Earth's surface. It is now clear that these volcanic chains are closely related to global tectonic activity. Many active volcanoes are located in the so-called "Ring of Fire" made up of island arcs and mountain ranges bordering the Pacific Ocean. The concept of seafloor spreading and, more broadly, the theory of plate tectonics offer a logical explanation for the location of most volcanoes.

Volcanoes Related to Plate Boundaries

Topographic maps reveal the locations of large earthquakes and indicate the boundaries of the 12 major tectonic plates. For example, the Pacific Plate is bounded by the earthquake zones of New Zealand, New Guinea, the Mariana Islands, Japan, Kamchatka, the Aleutian Islands, western North America, the East Pacific Rise, and the Pacific-Antarctic Ridge.

The Earth's plates, which move horizontally with respect to one another at a rate of a few centimetres per year, form three basic types of boundaries: convergent, divergent, and side-slipping. Japan and the Aleutian Islands are located on convergent boundaries where the Pacific Plate is moving beneath the adjacent continental plates—a process known as subduction. The San Andreas Fault system in California exemplifies a side-slipping boundary where the Pacific Plate is moving northwest relative to the North American Plate—a process called strike-slip, or transform, faulting. The East Pacific Rise is representative of a divergent boundary where the Pacific Plate and the Nazca Plate (west of South America) are moving apart—a process known as rifting.

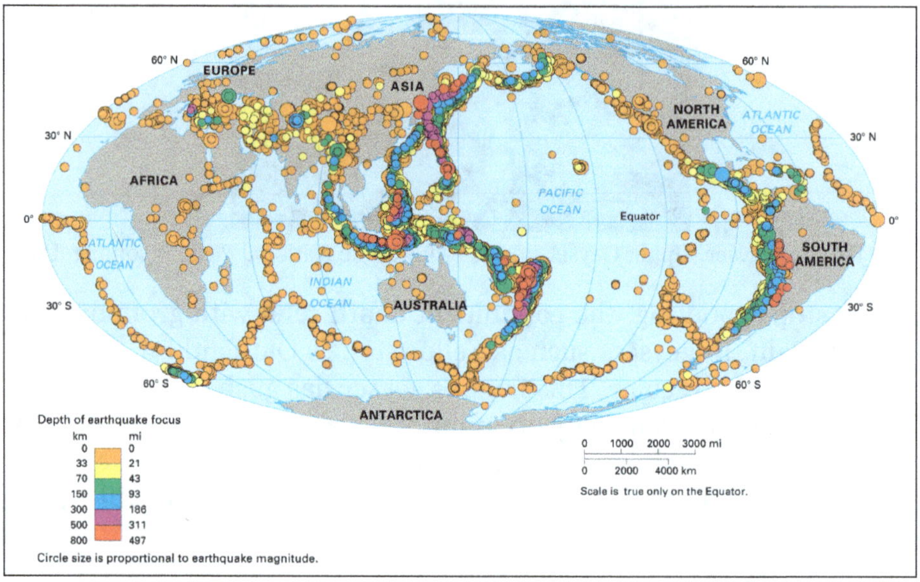

Volcanoes occur along both subduction and rift zones but are generally absent along strike-slip plate margins. Most subduction-related volcanoes are explosive and build stratovolcanoes, while rift volcanoes tend to be more effusive and build shield volcanoes, though there are exceptions to both these generalities. Subduction-related volcanoes erupt basalt, andesite, dacite, and rhyolite, andesite being the predominant rock type. Rift-related volcanoes, especially on the ocean floor, erupt mainly basalt.

The volcanoes on the western and northern margin of the Pacific Plate (New Zealand, New Guinea, Mariana Islands, Japan, Kamchatka, and the Aleutian Islands) are all subduction volcanoes. The rift volcanoes are largely hidden along the submarine crest of the East Pacific Rise and the Pacific-Antarctic Ridge at depths of 2 to 3 km (1.2 to 1.9 miles) below sea level. The Cascade volcanoes in the northwestern United States and the volcanoes in Mexico and Central America are related to the subduction under the North American Plate of the small Juan de Fuca and Cocos plates, which are on the east side of the Pacific Plate. Similarly, the volcanoes of the Andes are related to the subduction of the Nazca Plate beneath the South American Plate.

Conceptual models of how subduction and rift volcanoes may form are shown schematically in the diagram and in the video. Of the 1,450 volcanoes listed in the table of landform types, 80 percent occur along subduction zones, and 15 percent occur along rift zones. These percentages are somewhat misleading, however, because most of the

Earth's rift zones are about 2 to 3 km (1.2 to 1.9 miles) below sea level, where volcanic activity is hard to detect. At those depths active submarine volcanoes have yet to be observed, though many hydrothermal areas have been found along submarine rift zones by research submersibles. Iceland, a segment of the Mid-Atlantic Ridge that emerges above sea level, has 70 volcanoes that have erupted during the past 10,000 years. If this is a typical number for a rift system, there may be several thousand potentially active volcanoes along the oceanic ridges that are the surface expressions of the world rift system.

Subduction Volcanoes

As an oceanic plate is subducted beneath a continental plate, seafloor sediments rich in water and carbon dioxide are carried beneath the overriding plate. These compounds may act as fluxes, reducing the melting temperature of magma. Although the process is not clearly understood, magma apparently forms and rises by buoyancy from a depth of 100 to 200 km (60 to 120 miles). Subduction-zone volcanoes occur on the overriding plate and are offset inland from the actual plate boundary along the ocean trench.

The rising subduction-zone magma is probably basaltic in composition and is formed by the partial melting of mantle rocks. As the rising magma moves slowly up through the continental crust of the overriding plate, however, two things may occur to increase significantly the silica content of the magma. Crystallization of olivine and pyroxene minerals from the basalt can leave the residual melt enriched in silica and depleted in magnesium, iron, and calcium. This process is called fractional crystallization. Also, basaltic magmas have enough excess heat to partially melt the continental host rocks through which they are ascending. Because continental rocks are generally higher in silica, potassium, and sodium than are oceanic rocks, this process of assimilation and mixing can also play an important role in producing the wide range of compositions that occur in rocks from subduction volcanoes.

The additional gas content of many magmas at subduction volcanoes (which, coupled with their often high-viscosity magma, makes them dangerously explosive) may be explained by more than one process. Additional water and carbon dioxide may come from both subducted seafloor sediments and assimilated crustal rocks. Furthermore, any fractional crystallization tends to concentrate volatile elements in the residual melt. If volcanic gases form separate fluid phases within batches of ascending magma (as carbon dioxide gas is most likely to do), these fluid phases may ascend more rapidly than the overall magma body and be concentrated in the upper portion. Sudden expansion of these hot volcanic gases at atmospheric pressure is the apparent reason for the highly explosive nature of many subduction volcanoes.

Rift Volcanoes

Rift volcanoes form when magma rises into the gap between diverging plates. They thus occur at or near actual plate boundaries. Measurements in Iceland suggest that the separation of plates is a continuous process but that the fracturing is intermittent, analogous to a rubber band that is slowly stretched until it snaps. Earthquake swarms and volcanic eruptions occur when the stretching exceeds the strength of the near-surface rocks, which then fracture along steeply dipping cracks parallel to the rift. Basaltic magma rising along these fractures causes Icelandic-type fissure eruptions.

Rift volcanoes in continental locations such as the East African Rift System are more complex. Assimilation of continental crust apparently gives them some of the characteristics more generally associated with subduction volcanoes, such as having a wider range of rock types and explosive habits.

Intraplate Volcanism

The 5 percent of known volcanoes in the world that are not closely related to plate margins are generally regarded as intraplate, or "hot-spot," volcanoes. A hot spot is believed to be related to the rising of a deep-mantle plume, which is caused by very slow convection of highly viscous material in the Earth's mantle. As hot but solid mantle rock moves upward, partial melting may occur from the lowering of its pressure-dependent melting temperature. Where a lithospheric plate moves over a hot spot, a chain of volcanic islands may be created. As the plate moves, the older volcanoes are transported away from the magma source and become extinct. The younger, active volcanoes are clustered at the end of the chain over the hot spot. It is not known how a volcanic hot spot maintains its position for millions of years while a plate passes over it. Detailed seismic sounding of the mantle should increase the understanding of the mechanism controlling hot spots.

Hawaiian volcanoes are the best examples of hot-spot volcanoes. The five volcanoes that form the island of Hawaii at the southeast end of the Hawaiian chain are all less than one million years old. Two of these, Kilauea and Mauna Loa, are two of the most active volcanoes in the world. Northwestward along the Hawaiian chain each island is progressively older. The extinct volcano or volcanoes that formed the island of Kauai are about five million years old. Topographic maps show a major submarine continuation of the Hawaiian Ridge to the northwest of the Hawaiian Islands and then a dogleg bend into the Emperor Seamounts, which comprise an entirely submarine ridge continuing northward to the edge of the Pacific Plate.

Ages of rocks obtained by dredging and drilling the Emperor Seamounts indicate that they are a continuation of the Hawaiian chain and that the Hawaiian hot spot has been active for at least 80 million years. The Pacific Plate has moved over this centre of volcanism, first northward and later northwestward, at a rate of approximately 8 to 10 cm (3 to 4 inches) per year. The bend between the Emperor Seamounts and the Hawaiian Ridge occurred about 45 million years ago and indicates a significant shift in the direction of movement of the Pacific Plate.

Small, isolated intraplate volcanoes may not be produced by a hot spot but rather may be the result of deep fractures within the plates that allow pockets of magma to leak to the surface. These pockets originate in the low-velocity layer, so named because earthquake waves travel more slowly through the hot, plastically deforming rocks of this region than in the overlying rigid plates. The low-velocity layer begins about 50 to 150 km (30 to 90 miles) below the surface and extends to a depth of roughly 300 km (180 miles). A small amount (a few percent) of the low-velocity layer may be molten. Once a sufficient volume of magma forms in the subsurface, it tends to rise from its own buoyancy. Any fracture system at the plate margins or within the plates will facilitate this process.

Volcanoes and Geothermal Energy

Geothermal energy is plentiful, but geothermal power is not. Temperatures increase below the Earth's surface at a rate of about 30 °C per km in the first 10 km (roughly 90 °F per mile in the first 6 miles) below the surface. This internal heat of the Earth is an immense store of energy. In the upper 10 km of rock beneath the conterminous United States, it amounts to 3.3 × 1025 joules, or about 6,000 times the energy contained in the world's oil reserves. The problem in utilizing geothermal energy is extracting it.

The natural escape of the Earth's heat through its surface averages only 0.06 watt per square metre (0.006 watt per square foot). To make geothermal power practical, some special situation must exist to concentrate the Earth's heat energy in a small area. Underground reservoirs of steam or hot water that can be funneled into a drill hole provide this special situation. Some geothermal steam wells can produce 25 megawatts of thermal power, an amount equal to the normal heat flux of more than 400 square km (150 square miles) of land surface. The key to this concentration is the transfer of heat from deeper levels to the near surface by the ascending magma associated with volcanism. Magma at temperatures close to 1,200 °C (2,200 °F) moves upward to depths of only a few kilometres, where it transfers heat by conduction to groundwater. The groundwater then circulates by convection and forms large underground reservoirs of hot water and steam. Some of this thermal water may escape to the surface as hot springs or geysers.

Holes drilled into a subsurface geothermal system allow rapid transfer of hot water or steam to the surface. At the Geysers, a geothermal field north of San Francisco, superheated steam is directly tapped from porous underground reservoirs. In most other geothermal fields, the hot water is at or below its subsurface boiling temperature—about 300 °C (570 °F) at a depth of 1 km (0.6 mile). The hot water and steam produced from geothermal wells are used as the energy source to drive turbine generators in electric power plants. Hot water from lower-temperature geothermal reservoirs can be used for space heating and other applications. This form of geothermal power is utilized extensively in Iceland.

Some geothermal systems act as natural distilleries in the subsurface, dissolving trace amounts of gold, silver, and other rare elements from their host rocks. These elements may then be deposited at places where changes in temperature, pressure, or composition favour precipitation. Many hydrothermal ore deposits have been formed by once active—and in a few cases still active—geothermal systems. Gold is one more legacy of volcanism.

Magma

Magma is a completely or partially molten natural substance, which on cooling, solidifies as a crystalline or glassy igneous rock. It is usually rich in silica and capable of flowing under moderate differential stress. Magmas may carry rock fragments or crystals in suspension, and they normally contain gaseous (volatile) components in solution.

Volcanic magmas fall within a strictly limited compositional range that reflects the physical and chemical processes responsible for their generation and differentiation. Our concern is the physical phenomena of volcanism, interpretation of which requires some knowledge of physical properties of magmas.

Unfortunately, we have only a meager knowledge of liquid properties. Much of what is known can be explained in terms of the properties of Silicon (Si) and Oxygen (O) ions, which are usually the most abundant components. Si has a high charge (+4), small ionic radius (0.39 Å), and low coordination number with oxygen (4 oxygens surround each silicon, forming the corners of a tetrahedron). This results in strong ionic field strength and bonding with oxygen compared to other cations: Ca, Mg, Fe, Mn, Ti, Na or K. Al, which has similar but not as strong properties, plays a similar role to Si in both liquids and crystalline solids.

Structural State of Silicate Melts

Modern concepts of silicate liquid structure are based on the Zachariasen Model. The atoms are bonded by forces similar to those between atoms of crystals, but lack long range periodicity and symmetry. The magmas have silica (and alumina) tetrahedra linked (or polymerized) in threedimensional networks in which (bridging) oxygen atoms are shared by two or more tetrahedra; the Si and Al cations are termed "framework cations." Other cations enter the melt in limited amounts as independent ions occupying positions between tetrahedra, and modify the basic structural framework and its physical properties; these cations, Ca, Mg, Fe, Mn, Ti, Na, and K, are termed "framework-modifying cations."

The framework-modifying cations can be accommodated in amounts of up to about 20 cation percent before the basic framework breaks down into smaller geometric units. In breaking liquid continuity into smaller units, the framework changes from an extensive network of tetrahedra, all of which are linked by shared O atoms to smaller units with lower Si:O ratios until, when more than 66% of the cations are framework modifiers, the liquid consists of separate tetrahedra not directly linked to each other.

Melt structure controls the physical properties of a magma. Viscosity is the most important of these properties, because it plays a role in factors controlling both the style of volcanic eruption and the physical nature of volcanic products.

Viscosity

Viscosity is a fluid's internal resistance to flow. It represents the ratio of shear stress to rate of shear strain applied to a layer of thickness Z and permanently deformed in a direction x parallel to the stress. Mathematically, viscosity is expressed by:

$$s = s_o + \eta \left[\frac{dm}{dt} \right]^n$$

where s is the total shear stress applied parallel to the direction of deformation; s_o is the yield strength of the fluid or the stress required to initiate flow; ç is the viscosity, expressed in units called poises (dyne sec/cm2); dm/dt is the gradient of velocity dx/dt or strain rate over a distance Z normal to the direction of shear; and, n is an exponent which has a value of 1.0 or less depending on the form of the velocity gradient.

For many fluids, this expression describes a linear relation between the strain rate (dx/dt) and shear stress parallel to the direction of shear. If a shear stress greater than the yield strength ($s > s_o$) is applied, the resulting strain has two components:

1. Elastic and recoverable; and,
2. Viscous and non-recoverable.

If a stress less than yield strength ($s < s_o$) is applied, the substance is deformed elastically and returns to its original form after the stress is removed. Some fluids do not require application of some initial force before they are permanently deformed by shear stress parallel to the direction of shear. Such fluids are said to exhibit Newtonian behavior when n equals 1.0 and s_o equals zero.

Highly polymerized or non-Newtonian fluids (known as Bingham liquids) have a finite yield strength that must be exceeded before they can be deformed permanently. In other words, Bingham fluids behave elastically until their yield strength is exceeded.

Cooling and crystallizing magmatic liquids behave as newtonian fluids only until they contain approximately 20% crystals. Liquids with suspended solid particles may have a non-linear relation of shear stress to strain rate, for which the value of n is less than 1.0.

Controls on Viscosity

Various factors control magmatic liquid viscosity: composition (especially Si and volatiles), temperature, time and pressure, each of which effect the melt structure. Actually, the viscous behavior of complex silicate liquids, such as magmas, is difficult to predict, because no comprehensive theory explains the effects of major cations or temperatures of magmatic conditions.

It is possible to estimate the viscosity of a magmatic liquid at temperatures well above liquidus temperatures (that is, temperatures at which only liquid is present) from chemical compositions and empirical extrapolation of experimental data on the linear relationship between h and temperature in simple chemical systems. The range of temperatures of naturally flowing magmas, however, is near or within the crystallization interval, where stress-strain relationships are not linear (that is, they are crystal-liquid mixtures and show Bingham behavior). Under such conditions, the only way to predict viscosities is by analogy with similar compositions investigated experimentally.

Silica Composition

The strong dependence of viscosity of molten silicates on Si content can be illustrated by those of various Na-Si-O compounds:

Na:Si:O	η (poises)
0:1:2	10^{10}
1:1:2.5	28
2:1:3	1.5
4:1:4	0.2

The decrease in viscosity can be attributed to a reduction in the proportion of framework silica tetrahedral, and therefore, strong Si-O bonds in the magma.

Temperature

Temperatures of erupting magmas normally fall between 700° and 1200°C; lower values, observed in partly crystallized lavas, probably correspond to the limiting conditions under which magmas flow. Low temperatures characterize silica-rich rhyolite magmas, whereas the highest temperatures are observed in basalts. Magmas do not crystallize instantaneously, but over an interval of temperature. Few magmas, however, have a wide enough range of crystallization to remain mobile at temperatures far below those at which they begin to crystallize or much hotter than those temperatures.

Temperature has a strong influence on viscosity: as temperature increases, viscosity decreases, an effect particularly evident in the behavior of lava flows. As lavas flow away from their source or vent, they lose heat by radiation and conduction, so that their viscosity steadily increases. For example:

1. Measured viscosity of a Mauna Loa flow increased 2-fold over a 12-mile distance from vent;

2. Measured viscosity of a small flow from Mt. Etna increased 375-fold in a distance of about 1500 feet.

The decrease in viscosity can be attributed to an increase in distance between cations and anions, and therefore, a decrease in Si-O bond strength.

Time

At temperatures below the beginning of crystallization, viscosity also increases with time. If magma is undisturbed at a constant temperature, its viscosity may continue to increase for many hours before it reaches a steady value. The viscosity increases with time results partly an increasing proportion of crystals (which raise the effective magma viscosity by their interference in melt flow), and partly from increasing ordering and polymerizing (linking) of the framework tetrahedra.

Volatiles

The solubility of gases in magmas varies with pressure, temperature and composition of both the gas and the magmatic liquid. Because the volume of a melt with dissolved gas is less than that of a melt and separate gas (vapor) phase, solubility increases as gas pressure increases. At constant gas pressure less than total pressure, any increased load pressure on the melt lowers solubility, because the volume of the melt with dissolved gas is greater than that of melt alone.

Vapor pressure increases with temperature, so that solubility of any volatile component generally decreases with temperature, except possibly at high pressure. Consequently it is difficult to predict how volatile content of magma varies with depth. Nevertheless, it has been shown that at constant temperature, solubilities of water in magmas with different compositions are not significantly different.

Nearly all magmas can contain more water or gases at depth than they can continue to hold in solution when they reach the surface. Basalts, however, normally contain less water than rhyolites simply because their temperatures are higher, and thus, as noted, lower gas solubility. Only limited data exists concerning the effect of volatiles (in particular F, Cl, S, H_2S, SO_2, CO, and CO_2) on magma viscosity. No doubt, the effect of dissolved water is to lower viscosity, the effect being greater for silica-rich than silica-poor magmas:

Magma	T(°C)	η_{dry}(Poises)	η_{wet}(Poises)
Rhyolite(70% SiO_2)	785	10^{12}	10^6(5% H_2O)
Andesite(58% SiO_2)	1000	10^4	$10^{3.5}$(4% H_2O)
Basalt(48% SiO_2)	1250	10^2	10^2(4% H_2O)

Dissolved water disrupts the framework of linked Si and Al tetrahedra, but where such polymerization is already minor or absent, there is little effect. F and Cl are though to considerably reduce magma viscosities; in contrast, CO_2 increases polymerization, and therefore viscosity, in melts by forming CO_3^{2-} complexes.

Pressure

The effect of pressure is relatively unknown, but viscosity appears to decrease with increasing pressure at least at temperatures above the liquidus. As pressure increases at constant temperature, the rate at which viscosity decreases is less in basaltic magma than that in andesitic magma. The viscosity decrease may be related to a change in the coordination number of Al from 4 to 6 in the melt, thereby reducing the number of framework-forming tetrahedra.

Crystal Content

The effect of suspended crystals is to increase the effective or bulk viscosity of the magma. The effective viscosity can by estimated from the Einstein-Roscoe equation:

$$\eta = \eta_o(1 - RC)^{-2.5}$$

where η is the effective viscosity of a magmatic liquid, C is the volume fraction of suspended solids; η_o ho is the viscosity of the magmatic liquid alone; and, R is a constant with a best-estimated value of 1.67.

Bubble Content

The effect of gas bubbles (vesicles) on the bulk viscosity of magmas can be variable, and depends on:

1. The Degree of bubble formation (that is, vesiculation);
2. The size and distribution of bubbles;
3. The viscosity of the intervening melt.

Exsolution of water increases viscosity, but the exsolved vapor is a very low viscosity fluid; in basaltic magmas, the bubbles may enhance the already low temperature and composition controlled viscosity. Rhyolitic magmas have high viscosities irrespective of the degree of vesiculation, and only effect of high bubble content will be to reduce mechanical strength of the melt.

Yield Strength

Most magmas have an appreciable yield strength, which shows a marked increase below their liquidus temperature. As yield strength increases, the stress required to initiate and sustain flow becomes greater, and the magma's apparent or effective viscosity is also increased.

Specific Heat

The specific heat (Cp) of magma, which is the heat required to change the temperature of the liquid 1 degree Celsius, is typically about 0.3 cal. gm^{-1}. The specific heat contrasts greatly with heat of fusion or crystallization, which is the heat that must be added to melt or removed to crystallize a unit mass that is already at a temperature where liquid and solid coexist. Heats of fusion are typically about 65-100 cal. gm-1 at 1 atmosphere. Consequently, about the same amount of heat is involved in crossing the crystallization interval, as in raising or lowering the temperature of the rock or liquid through 300°.

Thermal Conductivity

Igneous rocks and liquids are poor conductors of heat. Thermal conductivity depends on two heat transfer mechanisms:

1. Ordinary lattice or phonon conduction;
2. Radiative or photon conduction.

The former declines and the latter increases as temperature increases and the melt structure expands. For rocks, the two effects balance each other up to their melting range. At high temperatures, the thermal conductivity of mafic rocks normally declines at an increasing rate up to 1200°C, above which, radiative heat transfer increases as does total thermal conductivity. More silica-rich rocks show increasing thermal conductivity at lower temperatures.

Density

Magma densities range from about 2.2 gm cm-3 for rhyolite to 2.8 gm cm-3 for basalts, illustrating a close density-melt composition relationship, primarily reflecting the influence of higher concentrations of Fe, Mg and Ca cations in basalts. In contrast, magma density decreases with increasing temperature and gas content. These densities increase a few percent between liquid and crystalline states.

The temperature dependence of magma density is given by the coefficient of thermal expansion, about 2-3 x 10-5 deg^{-1} for all compositions. The pressure dependence of magma density is given the compressibility or fractional volume change, $\Delta V/V$, per unit of pressure. Compressibility increases sharply in the melting range from 1.3×10^{-12} to about 7.0 x 10-12 cm^2 $dyne^{-1}$.

Electrical Conductivity

Electrical conductivity, which is low in pure silica melts, increases with increasing abundance of metallic cations, especially alkali elements, and increases abruptly in the melting range.

Seismic Wave Velocities

Compressional or P-wave velocities are about 6 km sec^{-1} up to the melting range, then decrease abruptly to 2.5 km sec^{-1} at higher temperatures. Shear or S-wave velocities are about 2-3 km sec^{-1}, which drop abruptly at melting temperatures.

Generation, Rise and Storage of Magma

The subsurface processes by which magmas are generated and rise toward the surface are extremely complex. Before examining these processes, it is worthwhile to review what is known concerning the Earth's interior.

Nature of Crust and Upper Mantle

Most of what is known concerning the Earth's interior comes from geophysical measurements, and concerns:

a. Seismic wave velocities;

b. Temperature;

c. Density distributions;

d. Heat flow;

e. Mechanical properties.

Seismic velocities increase with depth within the Earth, but show abrupt changes at several depths interpreted to represent discontinuities in the composition or structural state of minerals. The most notable discontinuities are:

a. Mohorovicic discontinuity (MOHO);

b. Low Velocity Zone (LVZ);

c. Core-Mantle boundary.

The seismic velocities are closely related to the density ρ and the elastic properties (bulk modulus K and rigidity or shear modulus μ by the following expressions:

$$V_p = \{[K + (4/3)\mu]\rho\}$$

$$V_s = (\mu/\rho)^{1/2}$$

The elastic properties are poorly known, but making certain assumptions, it appears that density increases to about 3.4 gm/cc at depths around 70 km, remains constant between 3.45 and 3.63 to the base of the Low Velocity Zone. Both pressure and temperature increase with depth. The temperature increase (6°/km) in the crust is consistent with an average heat flow of 1-2 x 10^{-6} cal. cm^{-2} sec^{-1}, with the highest values associated with young crust. If temperature gradients measured in the crust are projected downward, they rapidly approach temperatures for beginning of melting in the mantle near the Low Velocity Zone. The transmission of shear or S seismic waves, however, suggests the absence of large amounts of liquid, so that the temperature gradients must diminish with depth.

Heat Sources

Existence of magma indicates that at some depth beneath the Earth's surface, temperatures must be high enough to induce melting. One major problem associated with understanding the generation of magmas is the source of heat necessary to cause melt production. It is believed that the major source of heat within the Earth is the radiogenic elements, principally K, U, and Th. These elements, however, are concentrated within the Earth's crust, and have extremely low abundances in probable mantle rocks, too low to yield through their radioactive decay the heat necessary to generate magmas. Moreover, it can be shown that the melting process scavenges these elements, and thus, depletes even more their abundances in the source region.

Mechanisms of Melting

A variety of models have been invoked to explain the source of heat required to induce melting within the Earth:

a. Stress Relief: Pressure on the source region is released during tensional or compressional deformation of the overlying rock column.

b. Thermal Rise to Cusp in the Melting Curve: Intersection of pressure temperature conditions with the source rock melting curve under conditions where lowest temperatures on the solidus coincide with phase change boundaries.

c. Convective Rise : The source material rises by solid-state convection into a pressure-temperature regime appropriate for melting.

d. Perturbation: A local decrease in thermal conductivity or density leads to heating or diapiric rise of the source material.

e. Mechanical Energy Conversion to Heat: Force required to move one rock surface over another without grinding and deformation converted to heat, because of thrust faulting, subduction, a propagating crack or flaw in the Earth's lithosphere, shear or Tidal energy dissipated in the solid earth.

f. Compositional Change: The addition or subtraction of material changes the rock composition to a new composition whose solidus lies at a temperature less than the ambient temperature.

Partial Melting

Rocks are a heterogeneous assemblage of minerals, and each mineral is characterized by a unique melting temperature. Melting commences at grain boundaries, usually where three crystals of minerals with the lowest melting temperatures meet. As melting progresses, channelways develop between grains. Temperatures probably never are high enough to completely melt the source rock, and only part of or some of the minerals melt. This process is therefore called partial melting.

Because of mechanical constraints, it is generally believed that at least 1-5% melting is required for the melt to separate from the unmelted (refractory) solid (crystalline) material. Melting probably never exceeds 35% because of the gravitational instability of low density liquid with higher density refractory minerals. The composition of a partial melt (magma) depends on the melting conditions present in the Earth:

a. Temperature;

b. Pressure;

c. Volatile content;

d. Mineral composition of the source rock;

e. Amount or degree of melting.

Once gravitational instability sets in, the melt separates from the solid (denser) residuum. Depending upon where separation occurs, the magma may ascend through ductile (mantle) and/or brittle (crust) domains within the Earth. The manner in which magma rises differs between these two domains.

Segregation and Rise of Magmas through the Mantle

Several mechanisms of magma rise through the mantle have been visualized. These processes include:

a. Deep Segregation: The melt forms along a dendritic network of joints and fractures in the zone of melting, and feeds into a smaller number of layer tributaries eventually forming a larger channel at higher levels. With melting concentrated along grain boundaries, melt migration is caused by a thermal or pressure gradient or by capillary effects. This migration the presence of a critical proportion of melt before solid/liquid separation occurs. Two factors which could provide the driving force following initial separation are:

- Pressure resulting from volumetric expansion on melting,
- The buoyancy of the liquid.

Once the liquid has separated, it is unlikely that it maintains a temperature much higher than its surroundings, as it is cooled by adiabatic expansion and conduction to the wall rocks. If the liquid rises slowly through rocks that are below their melting temperature, the magma would crystallize quickly. Thus, magmas can only ascend once the temperature of their wall rocks have been elevated, and successive batches of magma must tend to follow paths of earlier bodies.

b. Diapiric Rise: A density reversal can lead to what is known as Rayleigh Taylor instability in which lighter underlying material first collects in localized bulges under the heavier layer. The low density layer moves upward at an accelerated rate until it forms a steep sided plume or vertical density current. The rate of ascent , size, and spacing of plumes is a function of density differences, and the viscosity of the overlying rocks. Little or no separation of melt occurs in the zone of melting. Instead, the crystal-liquid mush rises and separation occurs at shallow levels. There again must be a delicate thermal balance between the diapir and its surroundings. Otherwise, it crystallizes.

c. Zone Melting: A body of magma rises by melting its roof, while it crystallizes on its floor. The zone of melting rises without actual movement of liquid and with little loss of heat. Heat used in melting is regenerated by release of latent heat of crystallization. It has been estimated that a body of magma 7 km thick starting at a depth could rise to within 8 km of the surface before crystallizing in about 1 million years.

Rise of Magmas through Brittle Lithosphere

It is difficult to determine the level at which the lithosphere deforms by brittle fracture rather than by plastic flow - a depth represented by earthquake foci. There is strong evidence, in the form of individual and swarms of dikes, that large bodies of magma are tapped within the crust at a level where rocks can fail by dilational fracture. However, temperatures and pressures in the vicinity of large magma bodies are not normally consistent with purely brittle fracture. The manner in which magmas rise through the lithosphere may be:

Dilational Rise: This proposed mechanism by which magma may rise involves: entrance of melt in fractures, and rise due to gravitational buoyancy; The fracture becomes extended vertically and/or horizontally along a plane normal to the minimum stress; and, the fracture closes behind the magma as it passes and pressure on the wall falls below the confining pressure, rebounding due to viscoelastic deformation. Such a mechanism may explain the limited duration of basaltic fissure eruptions and the apparent arrival of discrete batches. Many instances, however, exist where acid or volatile magmas have apparently risen as pipe-like intrusions with little or no evidence of horizontal deformation.

The ability of a magma to rise through brittle lithosphere is usually explained in terms of depth and density contrast with the overlying rocks. If the pressure on the magma is equal to the lithostatic load of overlying rocks, the magma can rise to a level determined by the density contrast. At a depth of 50 km, the lithostatic pressure can exceed the pressure of a vertical magma column enough to segregate liquid and cause it to rise. If the heights to which magmas can rise is solely dependent on the depth to source and a density equilibrium, it would be expected that magmas with deep sources would erupt at higher elevations, and vice versa. This is obviously not the case as demonstrated by volcanoes of the Mexican volcanic belt.

More important limitations to magma rise are probably the heat content, and rates of ascent and cooling, which in turn, depend on the size of the magma body. Another important factor is the stress regime, which governs the form of the intrusive bodies. The three basic magma stress regimes are:

a. Least principal stress is horizontal (dikes);

b. Least principal stress is vertical (sills);

c. The stresses (vertical and horizontal) are equal (pipes; random dikes and sills).

At relatively high magmatic pressures or at shallow depths where vertical and horizontal stresses are low and about equal on the surrounding rocks, the magma conduits tend to be cylindrical. Thus, the form taken by a magma body may change drastically during its ascent. It is likely that near the surface, a cylindrical pipe is the most efficient form of conduit, because flow velocity increases and heat losses decrease as the horizontal section increases in size and becomes equidimensional. Thus, conduits tend to become centralized at the intersection of two or more fracture systems.

Non-Dilational Rise: As mentioned previously, there is ample evidence that some magmas have forcibly displaced rocks into which they have intruded, but others have made room for themselves by stoping or elevating the roof rocks. It is obvious that the critical elements are heat, and the manner in which the magma crystallizes, the shape and size of the body, and the volatile content of the magma.

An excellent example of non-dilational rise is illustrated by the formation of diatremes, steep-sided, more or less cylindrical or funnelshaped breccia pipes formed by penetration of crust by moderate temperature, gas-rich magma (kimberlite and carbonatite). Two mechanisms may be capable of boring through the Earth's crust and creating diatremes:

a. Highly energized gases of deep-seated origin bore through the crust, opening channelways for the rapid ascent of magma;

b. Explosive eruption is triggered by vaporization of heated groundwater propagated downward as pressure is released on progressively deeper gas-charged horizons.

Flow of Magma

Knowing the rheological or fluid properties of magmas, we might be able to apply basic fluid dynamic principal to predict flow regimes of intrusive and extrusive magmas under various physical conditions. Unfortunately, a rigorous approach to our understanding of flow characteristics is not currently possible in the face of incomplete information about essential parameters of specific cases. Nevertheless, some insight into magma ascent processes may be gained by considering simple examples and approximations.

Flow Rates

The volumetric flow rate of a viscous fluid through a cylindrical channel under a constant pressure gradient is given by:

$$Q = \left(P\Pi r^4\right) 8\eta L,$$

where Q is the volume flow rate in cm^3 sec^{-1}, P is the pressure drop in bars, r is the channel radius in cm, η is the viscosity of the fluid in poises, and L is the length of the channel in cm. Applying this relationship to a large (about 200 km^3) simple funnel-shaped magma chamber which is filled with basaltic magma (h = 300 poises) via a 3-km-long, 200-m-wide, cylindrical feeder pipe at its base and a pressure drop through the pipe of 1000 bars (1 kb/3.3 km), we find:

$$Q = \left(3.14 \times 10^{16}\right)\left(8 \times 300 \times 3 \times 10^5\right) = 4.36 \times 10^{10}\ \mathrm{cm}^3 / \mathrm{sec}$$

$$\text{or } 3.76 \mathrm{km}^3 / \mathrm{day}.$$

This simple calculation is important in that it illustrates that movement of large quantities of magma in short periods of time is entirely feasible.

Nature of Flow Regime

The type of flow imposed on a magma, that is, laminar or turbulent flow, is also of interest. For example, in the case of an initially heterogeneous magma, the liquid would become effectively homogenized by turbulence. The conditions that determine laminar or turbulent flow can be determined by calculating the dimensionless Reynolds number, Re, which in terms of average flow rate is given by:

$$Re = (2\rho Q)/\eta r \Pi,$$

where ρ is the density of the fluid. Turbulent flow occurs when Re > 2000. For the previous example, with ρ = 2.6 gm/cm^3,

$$Re = \left(2 \times 2.6 \times 4.36 \times 10^{10}\right) / \left(3.14 \times 10^4 \times 300\right) = 2.39 \times 10^4$$

Hence, flow of the basaltic magma within the conduit would be turbulent. The higher viscosity of acid magmas, however, renders turbulent flow unlikely in these cases. Because the viscosity of magmas normally exceeds 10^3 poises and velocities are rarely greater than a few cm/sec, flow is probably laminar under most geologic conditions.

It can be expected that the non-Newtonian characteristics of magma also have an effect on flow behavior. Because a certain yield strength must be exceeded before many magmas can be deformed by viscous flow, velocity gradients in the margins of a moving magma are likely different from those of more familiar liquids like water.

Shear stress in the boundary of the moving liquid is greatest near a stationary surface and diminishes toward the interior. Thus, if viscosity is uniform throughout the entire flow width, then the velocity distribution is parabolic. But if heat is lost at the stationary boundary and the effective viscosity increases sharply with falling temperature, the flow profile is more arcuate. These different flow profiles reflect both the effect of falling temperature on both viscosity and yield strength of the magma.

In many cases, it is likely that a zone of static liquid will form a layer between the moving liquid and its solid boundary. Heat transferred from a cooling magma to surrounding wall rocks also affects its behavior in other ways.

Flow Instabilities

When heat losses from the top or sides of a magmatic body cause a density difference in the liquid large enough to produce gravitational instability, the liquid overturns and free convection accelerates the rate of heat transfer. The onset of convection in an infinite horizontal layer of viscous fluid having an upper and lower surface is given by the dimensionless ratio of buoyant to viscous forces known as the Rayleigh number, Ra:

$$Ra = (L^4 \alpha_T g \beta) \eta K$$

where L is the height of the layer in cm, α_T is the coefficient of thermal expansion, g is the constant of gravitational acceleration (980 cm/sec), β is the vertical temperature gradient in K cm^{-1}, η is the kinematic viscosity (η / ρ), K is the thermal conductivity of the magma in cal gm^{-1} K^{-1}, and ρ is the fluid density. Ra for a vertical tube heated from below is given by the same expression, except that L^4 is substituted by r^4 where r is the characteristic radius of the tube in cm.

The critical Ra value above which convection begins is about 1700, approximately the same value calculated for magmatic bodies of most common shapes. For a magma body of given size and viscosity, the principal variable is thermal gradient, β, a function of heat loss to the top or sides of the magma body. For Ra < 1000, transfer of heat is predominantly by conduction; steady convective heat transfer sets in at approximately Ra > 10000, and strong eddying motion is attained when Ra = 100000.

Bodies with thickness or radius greater than 10 m are likely to convect if their heat losses are those that would be expected at shallow crustal depths (10^{-5} to 10^{-3} cal cm^{-2} sec^{-1}). Clearly, the larger the magma and the lower its viscosity, the more likely convection occurs, but quite small bodies having high heat flux values, should also be quite unstable.

High-level Reservoirs and Subvolcanic Stocks

The erosion of extinct volcanoes reveals the presence of simple and multiple stocks of medium- to coarse-grained rocks. Generally, the stocks are 1- to 10-km-wide, circular to oval in cross-section, and grade upward into a maze of inward dipping sills, steep radial dikes, and cone sheets. Most of these intrusive rocks have made room for themselves by stoping rather than forcible intrusion. There is good evidence that these intrusive bodies were volcanic reservoirs, because compositional features of erupted materials indicate that most magmas tended to reside and equilibrate in such shallow reservoirs prior to eruption. Other than what we see within deeply eroded volcanoes, however, little is known concerning the volcanic reservoirs beneath active volcanoes, except what is indicated by geophysics:

1. Seismic methods: These methods have been used to detect large magma bodies at depth because of the inability of the Shear or S seismic waves to be transmitted through liquids. The distribution of earthquakes generated within or directly below a volcanic structure may delineate:

   - The Boundaries of intrusive bodies, and
   - The Possible movement of magma within the subvolcanic plumbing system.

   For example, a three-dimensional distribution of earthquake foci surrounds an aseismic zone, which may represent one or more bodies of magma beneath Kilauea. Several types of earthquakes of volcanic origin are recognized according to the location of their foci and the nature of earthquake motion:

   - A-type volcanic earthquakes: These earthquakes take place in and beneath volcanoes at places deeper than 1 km, generally in the range from 1 km to 20 km. They are generally less than 6 in magnitude.
   - B-type volcanic earthquakes: These earthquakes originate usually in and adjacent to active craters at extremely shallow depths. The magnitudes are generally extremely small. The earthquake motions consist mainly of vibrations with periods in the range of 0.2 sec. to 1.0 sec.
   - Explosion earthquakes: The maximum amplitude or magnitude of the earthquake has a close relationship with the intensity of explosive eruption and is approximately proportional to the kinetic energy of the eruption. The earthquake motions show a predominance of longer wave length as

compared with those of the A-type volcanic and tectonic quakes. The associated detonations or air vibrations of explosive eruptions are remarkably strong.

- Volcanic tremors: Earthquakes take place incessantly or continuously with a short interval, such as every several seconds, so that motions are recorded continuously. These earthquakes may originate from extremely shallow positions in or near the crater, or at deep levels (20-30 km at Kilauea). Various wave forms are found in volcanic tremors, including surface waves of Rayleigh and Love type.

2. Gravity Measurements: Precise gravity measurements may also reveal the presence of an anomalous mass of magma at depth, and provide a means of constructing subsurface structural models. Gravity surveys have shown that the Hawaiian volcanoes have crudely cylindrical cores composed of dense rock only a few km below their summits. Gravity measurements have also suggested the presence of large batholith-size, low-density bodies of magma or intrusive rock beneath many large calderas. They also indicate that Cascade volcanoes lie within grabens, or down-dropped tectonic blocks, underlain by similar subvolcanic intrusions.

3. Infrared Radiometry: This technique is used to detect the presence of bodies of rock or magma at elevated temperatures.

4. Tiltmeter Measurements: Precise leveling and tilt measurements have been used to detect deformation caused by the intrusion of magma into shallow levels. Such measurements have been used to estimate the depth and geometry of the intrusions, because they provide precise information concerning the horizontal as well as the vertical components of movement.

Lava

Lava is magma (molten rock) emerging as a liquid onto Earth's surface. The term lava is also used for the solidified rock formed by the cooling of a molten lava flow. The temperatures of molten lava range from about 700 to 1,200 °C (1,300 to 2,200 °F). The material can be very fluid, flowing almost like syrup, or it can be extremely stiff, scarcely flowing at all. The higher the lava's silica content, the higher its viscosity.

Mafic (ferromagnesian, dark-coloured) lavas such as basalt characteristically form flows known by the Hawaiian names pahoehoe and aa (or a'a). Pahoehoe lava flows are characterized by smooth, gently undulating, or broadly hummocky surfaces. The liquid lava flowing beneath a thin, still-plastic crust drags and wrinkles it into tapestry-like folds and rolls resembling twisted rope. Pahoehoe lava flows are fed almost wholly internally by streams of liquid lava flowing beneath a solidified or partly solidified surface. Typically, the margin of a pahoehoe flow advances by protruding one small toe or lobe after another.

Réunion, Volcano erupting on the island of Réunion, western Indian Ocean.

In contrast to pahoehoe, the surface of aa lava is exceedingly rough, covered with a layer of partly loose, very irregular fragments commonly called clinkers. Aa lava flows are fed principally by rivers of liquid lava flowing in open channels. Typically, such a feeding river forms a narrow band that is 8 to 15 metres (25 to 50 feet) wide along the centre line of the flow, with broad fields of less actively moving clinker on each side of it. At the front of the flow, clinkers from the top roll down and are overridden by the pasty centre layer, like a tread on an advancing bulldozer.

Pahoehoe lava flow, Kilauea volcano, Hawaii, November 1985.

Pahoehoe and aa flows from the same erupting vent are usually identical in chemical composition. In fact, it is common for a flow that leaves the vent as pahoehoe to change to aa as it progresses downslope. The greater the viscosity and the greater the stirring of the liquid (as by rapid flow down a steep slope), the greater the tendency for the material to change from pahoehoe to aa. The reverse change rarely occurs.

Lavas of andesitic or intermediate composition commonly form a somewhat different type of flow, known as a block lava flow. These resemble aa in having tops consisting largely of loose rubble, but the fragments are more regular in shape, most of them polygons with fairly smooth sides. Flows of more siliceous lava tend to be even more fragmental than block flows.

Thin basaltic lava flows generally contain many holes, or vesicles, left by bubbles of gas frozen into the congealing liquid. Thick flows, which remain hot for long periods, may lose most of their gas before the lava congeals, and the resulting rock may be dense with few vesicles.

Basaltic lava erupting from the Pu'u 'O'o spatter and cinder cone on Kilauea volcano, Hawaii.

Pyroclastic flows, which are low-viscosity, fluidized mixtures of hot but solid volcanic fragments and hot gas, are often described in newspaper accounts as lava flows. This causes much confusion. Molten lava flows are relatively high-viscosity liquids, and most of them advance slowly (a few metres per minute to less than a metre per day). Pyroclastic flows move more like a dense, low-viscosity gas pouring down a slope and even move upslope if they have enough momentum; their downslope velocities often exceed 100 km (60 miles) per hour.

Lava Flows

Lava flows are the most common volcanic feature on Earth. They provide fine construction materials and are also an essential source of nutrients for future agricultural soils. They nevertheless remain a persistent threat to human activity. After describing the primary features of lava flows, therefore, this chapter will focus on how such features can be used to improve forecasts of how a lava will behave.

The resilience, or perhaps stubborness, that keeps populations returning to hazardous districts is the chief reason why lava flows are a common threat to human settlements, even though they are produced by one of the least powerful styles of eruption. Between 1973 and 1998, lava effusions worldwide caused property losses of hundreds of millions of U.S. dollars (equivalent 1998 prices). Although these losses are small compared to those due to major explosive eruptions, they have long-lasting repercussions on local economies and are a key force driving investigations into how lavas behave and how defensive techniques can be improved.

A second motive for studying lavas is that they are the single most common feature on the surfaces of the terrestrial planets. They cover 90% of Venus, 50% of Mars, at least 20% of the Moon, and some 70% of the Earth, where they are mostly hidden from view on the ocean floor. Understanding how lavas are emplaced is thus crucial to reconstructing the surface evolution of the inner planets and to investigating the conditions below the surface that favour effusive eruptions.

The main factors controlling how lava flows develop are the rate at which lava is effused from the ground, the lava's physical properties, and the local environment (such as ground slope, topography and whether the eruption occurs on land, or below water or ice). Each of these factors can vary greatly among eruptions, as well as during a single effusion, and it might be expected that lavas should show a wide range of behaviour. In fact the contrary holds and common lava types evolve along only a few, clearly-defined trends, which link the morphology of lava surfaces to styles and rates of flow advance. Only by understanding why these natural evolutionary sequences occur will it be possible to improve strategies for mitigating lava hazard.

Lava flows are outpourings of molten rock, or magma. On Earth, the overwhelming majority have silicate compositions, for which common melting temperatures are in the range 800°- 1200°C; lavas of sulphur (e.g., Siretoko-Iosan volcano in Japan, and Lastarria volcano in Chile) and of carbonate compositions (e.g., Ol Doinyo Lengai volcano in Tanzania) also occur at lower temperatures (about 150°C for sulphur and 600°C for carbonatite), but these are extremely rare and are not important as far as general hazard studies are concerned.

Flows are distinguished from lava domes by their extreme elongation downslope. Historically, the volumes produced by single effusions of lava range from minor dribbles to outpourings of a few cubic kilometres (e.g., Etna, Sicily, 1614-24; Lanzarote, Canary Islands, 1730-36; Lakagigar, Iceland, 1783-85). The resulting flow fields can extend tens of kilometres, spread kilometres across and reach thicknesses of hundreds of metres, although most are more modest in size. The durations of single eruptions also cover a large range and, while some may reach decades, the majority lie between days and months. Rates of flow lengthening thus rarely exceed a brisk walking pace, so that it is usually possible for people to escape immediate danger. Exceptions occur during the start of effusions, when lavas can sometimes advance as fast as a galloping horse. In the saddest example, on Nyiragongo (Congo) at 10.15 in the morning of 10 January 1977, a fluid lava swept downslope at least 5 km in 20 minutes (15 km per hour), catching a small village unawares and roasting 70 people alive.

The Nyiragongo tragedy ghoulishly illustrates the need to prepare against lava invasion even before an eruption begins. To identify the most vulnerable districts, it is necessary to recognise probable locations of future eruptions and to forecast the likely travel distance of at least the initial lava flow. The first task is achieved by applying statistical analyses to the known distributions of vents at a volcano. The second requires a model which can link probable flow length to factors that can be measured before the next eruption begins, and it is here that the evolutionary sequences of lava flows assume a fundamental importance.

Evolutionary Sequences among Lava Flows

Most lavas are crystallizing upon eruption, owing to chemical imbalances induced in magma as it approaches the surface from below. They continue to solidify during effusion, aided by loss of heat to the ground and to the atmosphere. As a result, flows begin to form channels or tubes which concentrate motion along only a small number of paths, so that subsequent lava can be transported more efficiently from the vent to the front.

A flow initially forms a tube or a channel according to whether or not the lava surface can develop a continuous crust. When exposed to the atmosphere, a fresh lava surface chills to a strong, solid crust within minutes. At the same time, the new crust is pulled forward by more mobile lava. If the forward pull is large enough, the crust continually breaks into fragments and so, being unable to form a stable roof, the flow develops an open channel for containing the lava. If the forward pull is too small, a continuous crust can develop across the whole flow and this, anchored to the flow margins, gives birth to a tube.

A similar battle between crustal growth and disruption occurs at flow fronts. When disruption dominates, the front moves forward as a single unit, controlled by the properties of the frontal interior. When crust formation dominates, the front advances by oozing small tongues of lava through localised punctures in the crust. Whatever their style of motion, lava fronts are the slowest part of an advancing flow, in part because of crystallization, and in part because their areas of cross-section are larger than the active areas along feeding channels or tubes.

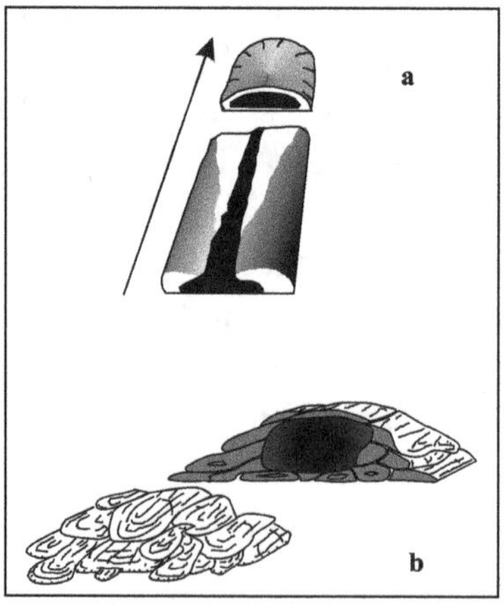

Figure above major flow structures: (a) In aa and blocky flows, open channels (below) typically feed lava to simple fronts (above). Motion, in direction of arrow, is concentrated in the black zones. For major flows, the fronts are commonly ~100 m wide and tens of metres thick. In long-lived eruptions, the channels may evolve into tubes. (b) The fronts of pahoehoe flows are normally a complex of intermingling tongues and toes (up to metres across and tens of metres long) fed by lava from a tube system.

From observation alone, it is clear that crustal structure can be linked directly to the early formation of lava channels and tubes, and to the style with which a flow front moves forward. The link with crustal structure is convenient since only the outer parts of an active flow are normally accessible to investigate, either by direct observation on the ground or by remote monitoring using aircraft or satellites. Difficulties in studying the rest of a flow arise because lavas are hot and viscous. To measure the temperatures, velocities and fluidities of flow interiors, it is necessary to stand close to poorly-crusted lava, where it is easiest to insert monitoring equipment. The most active parts of flows have mean surface temperatures commonly between 475°C (when their colour is a very dull red) and 1100°C (when they are golden yellow), a range sufficient for radiant heat to cause serious burns even metres away. Protective clothing, although cumbersome, can normally overcome this difficulty. However, even at the start of eruption, the most fluid lavas are frequently a million times more viscous than water and it is extremely difficult both to force a measuring device into a flow and, if successful, to retrieve it again.

In figure, (Left) a well-developed channel between static banks of aa lava. This example is a few metres wide, but channel widths may reach tens of metres in aa and hundreds of metres in blocky flows. Note the haze due to hot gases (mostly steam) escaping from the lava, moving from left to right. (Right) Extensive lava tubes are essential features of pahoehoe flows and may also form in aa flow fields during long-lived eruptions.

As a result, virtually no data are available from direct measurements of active lava interiors. Investigations must instead rely upon theoretical studies or upon indirect evidence from solidified lava interiors (exposed by natural collapse or by artificial excavations, such as road cuttings or quarries). In the first case, no theoretical model has yet been completely verified, precisely because the necessary field data are unavailable. In the second, the features preserved by solidified lava interiors (such as crystal size distributions, which depend on a lava's cooling history) may not reflect conditions that prevailed while the flow was still active, because these features (a) may have been dominated by changes (e.g., further crystallization) which occurred after the flow had come to rest, or (b) may have formed at different times during flow advance, and so cannot simply be related to the state of the active flow at any particular moment. Thus, until more reliable data are available for active flow interiors, the links between crustal structure and flow dynamics offer the best prospect of quantifying flow behaviour.

Table: Physical Properties of Lava Flows.

Composition	Eruption Temperatures °C	Density at Eruption Temperatures kg m^{-3} (without vesicles)	Viscosities at Eruption Temperatures Pa s*
Basalt Andesite Rhyolite Komatiite Water at Earth's surface	1050-1200 950-1170 700-900 >1600? 20	c. 2600-2800 c. 2450 c. 2200 c. 2800 c. 1000	$10^2 - 10^3$ 10^4-10^7 10^9-10^{13} <1? 10^{-3}
			Newtonian approximation at low shear rate
	Temperature °C (K)	Colour of Lava Surface	
	1150(>1423) 1090(1363) 900(1173) 700(973) 600(873) 475(748)	White Golden yellow Orange Bright cherry red Dull red Lowest visible red	

Field Classification of Lava Flows

Crustal appearance provides the basis for classifying lava flows on land into three major categories, pahoehoe, aa and blocky. Pahoehoe and aa are Hawaiian terms introduced in the late 19th century to describe the common lava types found on Mauna Loa and Kilauea, but they apply equally to other lavas with silica contents less than about 50-55 wt% (basalts and some basaltic andesites), as well as to the rare flows of sulphur and carbonatite. Blocky flows are common among lavas with silica contents greater than 55 wt% (basaltic andesites to rhyolites). Diagnostic features are most evident when viewing a crust over distances of decimetres and metres. At these distances, pahoehoe surfaces are smooth and, though occasionally broken, are normally continuous, while aa surfaces are extremely irregular, frequently fractured, and usually covered by rough, contorted fragments with typical dimensions of centimetres and decimetres. Blocky lavas, like aa, also have fractured surfaces and a covering of debris; they differ from aa flows in that their fragments are smooth and angular with common dimensions from decimetres to metres. From a distance, indeed, both aa and blocky surfaces look as exciting as piles of rubble on a building site.

Aa and Blocky Flows

Aa and blocky flows show simple evolutionary trends. As might be anticipated from their broken surfaces, their fronts tend to advance as single units and it is rare for one part of a front to move far ahead of neighbouring sections. Blocky fronts crumble to produce a snout of debris from the early stages of emplacement. Aa fronts show a greater range of behaviour, often starting as fluid sheets, but finishing as near-solid masses that fragment throughout their thickness to maintain advance; between these limits, they move by various combinations of fracture and flow.

Fronts thicken while advancing, often growing to more than ten times their initial thickness. Final thicknesses are typically about 20 m or less for aa fronts, but several tens of metres for blocky flows; their maximum lengths are measured in tens of kilometres and in kilometres, respectively. Major flows can achieve volumes of 1-100 million cubic metres, and tend to be emplaced within days when they are aa but within months when they are blocky.

Table: Common Features of Flow Surfaces.

Feature	Description
1. Aa Lava	Surface is covered by a jumble of irregular crustal fragments.
Cauliflower	Crust twists upwards as cauliflower-like protrusions. These break to give fragments up to decimetres across. Surfaces are grey-black, often glassy, and rough and spinose at the millimetre scale.
Rubbly	Crust fractures downwards to yield rounded rubble up to metres across, often with an ochre-black granular surface, millimetres deep
2. Blocky Lava	Surface is covered by broken lava, containing fragments up to metres across with smooth, planar, and angular surfaces.
3. Pahoehoe Lava	Surface is smooth and continuous, often with a millimetre-scale texture of interweaved lava threads or filaments.
Entrail	Dribbles of lava yield convoluted surfaces reminiscent of entrails.
Ropy (or corded)	Flexible crusts ruck into tight folds before chilling. Surface resembles segment of coiled rope. Each “rope” can be centimetres thick.
Shelly	Highly vesicular, fragile crusts. Often associated with skins, centimetres thick, over hollow lava blisters. The skins break underfoot, giving the impression of walking on egg shells.
Slabby (sometimes slab aa)	Slabs of broken crust, up to metres across and centimetres thick
4. Toothpaste Lava	Protrusions of viscous lava squeezed through gaps in flow crust. They may be tens of metres long and their cross-sections often mimic the shape of the source gap, like toothpaste emerging from its container.

Both flow types initially develop channels to feed lava to their fronts. Channels form when a flow stops widening and concentrates motion downhill. The fronts themselves grow during advance because they decelerate as they solidify and allow faster lava to accumulate from upstream. For example, during the opening stage of emplacement (when they are fastest), major aa fronts may advance a few kilometres a day (occasionally 10-30 km in the first 24 hours), although the velocity of lava near the vent may be at least ten times greater.

At one extreme, advance and thickening continue until a flow stops being fed by new lava, whereupon the front slows to a halt as remaining lava drains from the feeding channel.

At the other extreme, effusion continues into the flow even though the front has come to rest. Lava begins to pile up within the channel, starting at the front and working its way backwards. As it thickens, the channel lava exerts an increasing pressure on its margins, and these may eventually breach to form an outlet through which the channel lava can escape. If the breach is too small, it may be able to heal itself through cooling or by being plugged with crustal debris. Otherwise, the breach may become a permanent outlet from which a major new flow can develop. The new flow, in turn, may halt and thicken until the cycle is repeated. In such a way, a flow field, the final product of one effusive eruption, may evolve with time from a single flow to a collection of interconnected flows.

The surfaces of aa and blocky flows are covered by broken surface fragments. (Left) Aa fragments (Mt Etna, Sicily) are contorted and initially appear black and spinose (Left) but, during later stages of advance, the surface breaks to yield rounded and abraded rubble (Right). (Middle) Blocky fragments (Nea Kameni, Santorini, Greece) are angular and have planar faces. (Right) Pahoehoe surfaces (Vesuvius, Italy) are smooth and often billowy over distances of metres.

Breaching is more common among aa than blocky lavas. Apart from high channel pressures, the propagation of new flows by breaching requires channel lava that is much more fluid than its lateral margins. Only in this case can the channel lava easily escape through a breach; if, instead, the channel lava is almost as solid as the margins, breaching simply allows channel lava to spread into the breach and heal it. By virtue of its chemical composition, channel lava in aa flows tends to be more fluid than its blocky equivalent, so that conditions are more favourable in aa lavas for the propagation of new flows.

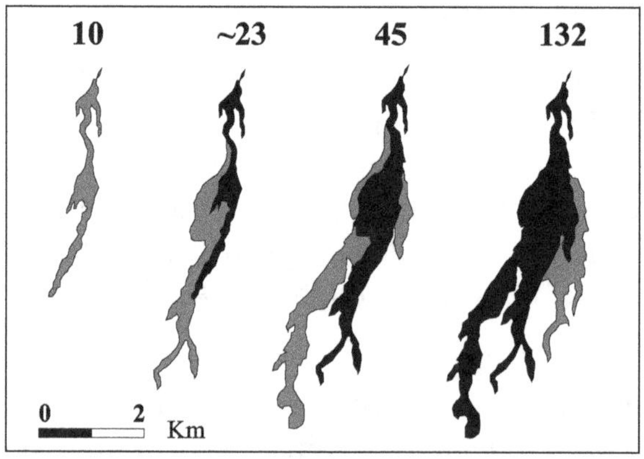

Although breaching may occur anywhere along a flow, it most commonly occurs somewhere along the upstream half of a flow's length. Newly-propagated flows may extend downslope beyond an earlier stream, but rarely do they increase the length of a whole flow field by more than half the length of the initial flow. A result of such behaviour is that the propagation of flows tends to widen, rather than to lengthen, the area covered by new lava. It also means that, when addressing the hazard from aa and blocky flows, the first goal is to estimate the probable maximum length of the initial flow.

As well as promoting flow breaching, flow thickening can trigger the overflow of fluid channel lava across its lateral margins. The margins can thus build themselves up as a series of superposed overflows which, under favourable conditions, develop an inward overhang across the channel surface. As the exposed channel surface narrows, it becomes easier for adjacent segments of crust to congeal together and to form a continuous roof over the flow. In this way, a lava channel can evolve with time into a tube, better insulating the lava beneath and allowing it to travel further before solidification sets in.

Aa flow fields grow as propagating a sequence of flows, new flows (grey) propagating from the sides of earlier streams (black). This example shows the evolution of Etna's 1983 flow field at 10, 23, 45 and 132 days after the start of effusion. Note how the final length of the flow field on Day 132 is only about 30% greater than the length of the initial flow on Day 10.

All flows initially lengthen so that their ratio of width to length decreases with time. This trend normally continues throughout eruption for pahoehoe and blocky flows. After the first flow is emplaced (arrow), aa flow fields tend instead to become wider. The initial flows rarely lengthen for more than 10-15 days. For all lava types, major flow fields are normally emplaced within months-years, although some historical effusions (especially of pahoehoe lava) have continued intermittently for a decade or more (e.g., Etna, 1614-24; Pu'u O'o, Kilauea, continuing since 1983).

Tubes form in aa and blocky flows only after a channel has become well-established, and may require several weeks to develop. When they occur, therefore, tubes are found along those parts of flows which remained active for long periods, typically behaving as feeders to sites of breaching downstream. As a result, tubes normally form after most major flows have been emplaced, and so rarely contribute in a significant way to extending a flow field (in complete contrast with pahoehoe flows, discussed below). As with the propagation of flows by breaching, the growth of lava tubes is favoured by a large contrast between the fluidity of internal lava and its external margins. Accordingly, tubes are more common in aa than in blocky flows. Should lava drain from a tube towards the end of effusion, it may leave behind a tunnel perhaps kilometres long that is large enough at least to crawl through. Among aa lavas, excellent examples can be found on Sicily's Mt Etna, including the 1971 and 1991-93 flow fields.

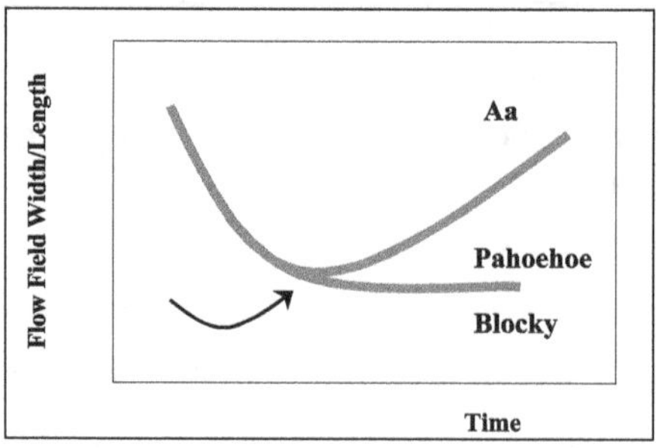

Pahoehoe Flows

Pahoehoe flows have dimensions similar to aa flows, but normally advance at least ten times more slowly. Because they spread slowly, their cooling surfaces resist extensive tearing and, though feeding channels can develop, it is not unusual for continuous crusts to form across a whole flow from the start of emplacement (unlike the weeks necessary in the case of aa lavas). The crust remains continuous around flow margins and fronts, whose initial thicknesses are usually of only a few decimetres. Since early flows are thin, they are easily retarded by the crust and spreading occurs by a combination of lava leaking out as small tongues through breaks in the crust, and by allowing new lava from upstream to burrow beneath the tongues and to slowly lift the front upwards. The front thus appears as a collection of intermingling tongues, each much narrower (by 10-1000 times) the width of the whole flow.

Pahoehoe tongues may extend by some 100 m, and even start to develop feeding channels, before crusting over and extruding small lava toes, typically a few metres or less in length (analogous to the lava pillows among submarine flows). As a result, pahoehoe fronts and margins soon develop as a complex of budding tongues and toes, all of which are gently uplifted by newly arriving lava. The surfaces of pahoehoe flows thus evolve a curious hummocky form, involving swells that extend distances from decimetres to hundreds of metres. The largest of these seem almost flat to the casual gaze; the smallest appear curiously grotesque and intestinal.

Lava from the vent continually raises the surface and, within weeks, a flow field may have thickened to several metres, while thin lava tongues continue to emerge from around its edges and from occasional surface ruptures. Although many tongues and toes stagnate after halting, others remain connected beneath the crust to form a network of distributary tubes, casually resembling an underground river system. Compared with the open channels in aa and blocky flows, lava tubes reduce the rate of lava cooling, especially when they are partly drained so that hot gases can collect beneath the tube roof and keep the flowing lava surface at temperatures close to its initial value. Thus, although pahoehoe flows advance more slowly than their aa counterparts, their interiors remain fluid for much longer periods. The greater lengthening time often dominates the slower velocity, so that pahoehoe flow fields can achieve lengths greater than aa flow fields of similar volume.

Historical pahoehoe flow fields have extended several tens of kilometres, the best examples being found on the Big Island of Hawaii. Some prehistoric flow fields, however, have been traced for more than 100 km, notably those in Queensland, Australia. Recently, it has been proposed that ancient flood basalts (such as the Columbia River Basalts in the U.S.A.), which have lengths of several hundreds of kilometres, are in fact enormous pahoehoe lavas, and not extreme aa flow fields as previously thought. The drained tube systems in such flows are truly impressive, reaching tens of metres across, up to 10-20 m high, and extending for several kilometres at least (the prehistoric Australian flow fields contain a tube system about 100 km long).

Lava Spectrum

In the simplest of cases, a flow maintains the same type of surface morphology throughout emplacement. Frequently, though, a flow surface evolves through more than one type with distance downstream. Among basaltic lavas, downstream transitions occur from pahoehoe to aa morphologies, while the change from aa to blocky is found among some basaltic andesites; both transitions are unidrectional, so that blocky surfaces do not become aa, and aa surfaces do not become pahoehoe. The pahoehoe, aa and blocky morphologies are thus not independent entities, but parts of a continuous spectrum of lava types. The association of the two transitions with different lava chemistry shows that composition is one underlying control. However, since each transition occurs between lavas of similar composition, other nonchemical factors must also be involved.

Both pahoehoe and aa surfaces are created during the formation of surface crust. At the start of eruption, the lavas are often too fluid to break before cooling, so that the degree of rupture must be controlled by the solidifying surface layers. To break a surface before it has chilled to its maximum strength, the rate of energy supplied predominantly by gravity to a unit volume of crust must be greater than a critical value. The critical value depends on how quickly a crack can be healed by chilling the newly-exposed lava beneath. If the critical value is exceeded, a flow breaks its crust more quickly than existing cracks can be healed and so evolves a fragmented aa surface; otherwise, a flow cannot tear its crust quickly enough and develops a continuous pahoehoe surface.

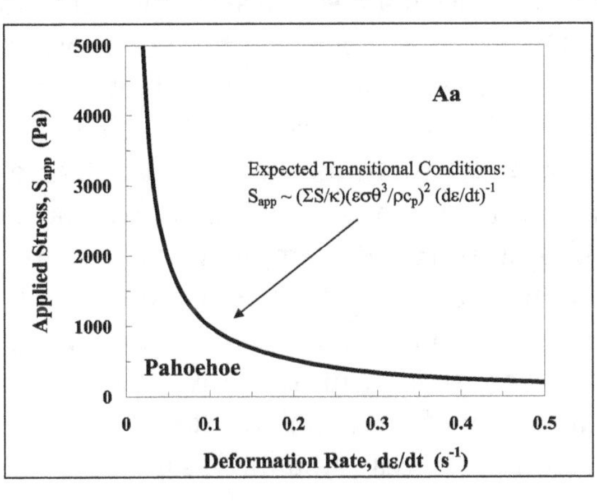

The evolution of pahoehoe and aa surfaces depends on the rate at which energy is supplied to deform cooling crust. The rate of energy supply per unit volume is given by the product of applied stress and deformation rate (Sapp dε/dt). Below a critical energy flux (~100 J m-3) controlled by the cooling and strengthening of lava exposed between cracks), the surface chills as a continuous crust to form pahoehoe. When this flux is exceeded, the surface breaks persistently to form aa. Blocky lavas are not shown on this diagram, because they are strong before eruption and, as if magmatic glaciers, they must break their surfaces to advance, independent of cooling effects.

The rate of energy supply is measured by the product of the stress applied to the crust and the rate at which the crust deforms. Two consequences are that the larger the pull, or applied stress, the smaller will be the deformation rate at which critical conditions are reached; and to cross the pahoehoe-aa transition, the rate of energy supplied (per unit volume) to the crust must also increase. The change from pahoehoe to aa occurs because, as lava advances downstream, a greater proportion of the gravitational energy flux for moving the whole flow is used in deforming a thickening crust. Eventually, the energy flux exceeds the critical value, after which the surface must break persistently as aa if the flow is to advance.

Blocky surfaces are associated with stronger and more viscous lavas that break before cooling is significant. In this regard, they can be considered magmatic glaciers, slowly-moving sheets that flow near their bases but break at their surfaces. The aa-blocky transition thus occurs when the lava interior becomes too crystalline to move forward only by flowing, so that surface fracturing is no longer controlled by crustal cooling. Indeed, the same reasoning explains why pahoehoe and blocky lavas have different surface morphologies, even though both advance much more slowly than aa flows; blocky flows are slow because of their high viscosity and break because of their large associated strength; pahoehoe flows are slow because they advance as thin sheets or tongues whose fluid interiors are easily restrained by surface crust.

The transitions between surface morphologies are irreversible; highly fragmented surfaces cannot recombine to form continuous crust (hence aa surfaces do not evolve into pahoehoe), while subcrustal lava does not become less solid with time (hence blocky surfaces do not evolve into aa). On occasion, however, it might appear that a reverse transition has taken place, especially between pahoehoe and aa types. The deception occurs under two main sets of conditions. The first corresponds to changing conditions of effusion from the vent. As an eruption decays to its close, so also decreases the rate of effusion. While early, fast lava may produce an aa surface, later lava may emerge slowly enough to form a continuous pahoehoe crust (good examples can be found on Mauna Loa, Hawaii). This change, though, does not represent a transition from aa to pahoehoe, since the later pahoehoe crust formed across an independent flow and did not evolve from lava already erupted with an aa surface. The second condition involves the escape from aa channels of hot internal lava, the result either of a channel overflow or of a margin being breached. In this case, the escaping lava may be less crystallized and, if advancing much more slowly than the parent flow, may form a pahoehoe crust. Once again, such a situation does not correspond to an evolution of the crust from aa to pahoehoe, because the escaping lava comes from deep within the parent flow and has had a crystallization history independent from that of the near-surface lava layers which formed the earlier crust.

Lava Dynamics and Internal Solidification

Because of its high viscosity, lava can rapidly reduce accelerations in a flow. As a result, flows tend to settle into a steady dynamic state. A simple example is shown by the preference of flow fronts to advance at nearly-constant velocities for extended periods of time. Since inertia is not important, flow growth is controlled by how a lava's rheological resistance is overcome by gravity (pulling lava downslope) and by pressure differences due to local variations in flow thickness (notably at the flow periphery).

Another consequence of low inertia is that lava strives to maintain laminar flow, whereby adjacent packets of lava tend to flow past each other, rather than intermingling, as would occur if motion were turbulent (although turbulence may be important in some exotic lavas). Without intermingling, a lava interior can diffuse heat only by conduction (i.e., interactions between neighbouring molecules), so that cooling effects tend to migrate inwards from the flow exterior, itself maintained at a low temperature by radiation to the atmosphere, by cooling wind, and by precipitation. Lava, however, is a very poor conductor. The time needed for cooling to penetrate a depth D into a flow is given approximately by D2 /4κ, where the thermal diffusivity (κ) of lava is between 10^{-7} and 10^{-6} m^2 s^{-1}. Thus it takes minutes to chill a layer centimetres deep, but weeks for this layer to thicken to metres.

To give an example of a lava flow 10 km long, the time for lava to travel from the vent to the front may be only hours for pahoehoe and aa flows, and days for blocky flows (Recall that this travel time is less than the interval since effusion began: the former is given by [flow length]/[mean velocity along a flow]; the latter is [flow length]/[mean velocity of the flow front].) The corresponding mean flow thicknesses will be measured in metres or tens of metres, implying conductive cooling times of weeks or longer. It might thus be expected that the interiors of pahoehoe and aa flow fronts will contain lava almost as fluid as it was near the vent. This condition is typical among pahoehoe flows, but not so for aa flows, which can develop solidified fronts within days.

Precisely why the interiors of aa flows can solidify so quickly is still not fully understood. One possibility is that, compared with other lava types, aa lavas are initially richer in volatiles. All magmas contain some volatiles, mostly water, that remain dissolved at depth, where the high pressure from surrounding rock prevents them from forming bubbles. As the magma ascends, the imposed pressure decreases and, eventually, bubbles can develop, just as they appear in a bottle of soda water when pressure is released as the bottle is opened. The loss of volatiles upsets the chemical balance in the liquid and can trigger crystallization at a rate that increases with the proportion of gases originally dissolved in the magma, but decreases with increases in liquid viscosity. Thus, compared with aa lavas, pahoehoe interiors may crystallize more slowly due to smaller amounts of initial gas, while crystallization in blocky interiors is retarded by high lava viscosity.

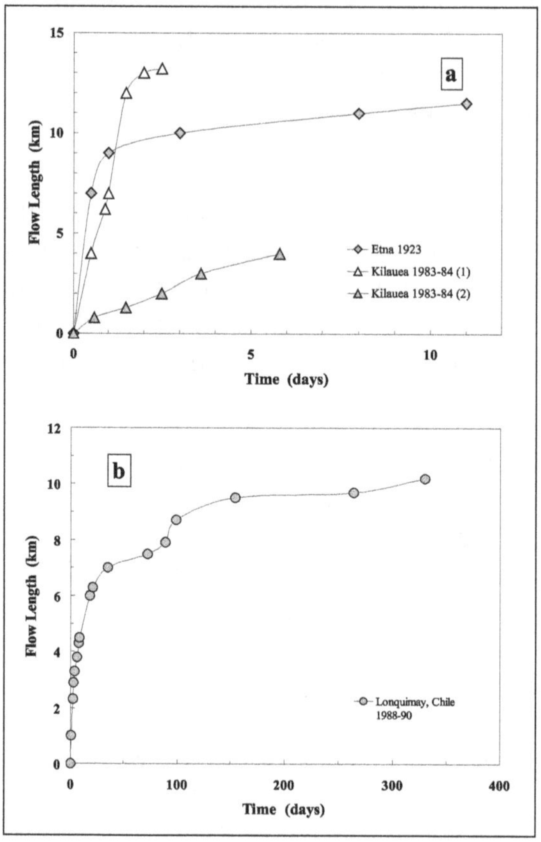

Most flow lengthening occurs during the early stages of effusion and at nearly constant rates of advance. This may be followed by a rapid drop in advance rate which, though continuing for a long time, does not extend a flow significantly.

Another explanation appeals to crustal entrainment. Pieces of cold crust are dragged back into the lava interior, accelerating solidification by forcing internal lava to intermingle. Importantly, the intermingling in aa is a result of crustal entrainment and not of natural turbulence; when possible, the flow again seeks a simple laminar motion. This mechanism is less common in pahoehoe flows, since they form continuous crusts, and also in blocky flows, because their very large viscosity impedes incorporation of surface debris.

Shapes of Lava Flow Margins

As they negotiate topographic irregularities and adjust to local variations in physical condition, lava flows are rarely able to spread evenly over the ground and so usually develop margins with irregular outlines. The degree of margin irregularity varies with flow type and, since all types can evolve over similar topography, such variation must reflect differences in style of advance and, hence, also in the governing force balance.

Spreading as collections of tongues and toes, pahoehoe flows develop lobate margins whose degree of unevenness (measuring fluctuations about a smoothed, average outline) happens to be similar for averaged outlines at least 1-100 m in length. Such a scale-independent (fractal) unevenness suggests that pahoehoe spreading is governed by the ratio (scale independent by definition) of imposed forces, and that these forces are of the same type at each scale. Since crustal restraint is essential to the pahoehoe style of advance, the obvious inference is that, whatever their size, tongues and toes propagate when, given sufficiently fluid internal lava, the forces driving lava spreading exceed crustal resistance by only a critical amount; otherwise spreading would cease (driving forces too small) or the crust would fragment and have no significant resistance (driving forces too large).

In contrast, aa and blocky flows spread mostly under conditions for which the driving forces are far greater than crustal resistance. Thus, although locally influenced by crustal debris, the bulk shape of their margins is more dependent on the rheological resistance of internal lava. As a result, the margins appear highly irregular over distances covering a few crustal fragments (notionally 10 m or less), but tend to show only subdued undulations over 100 m or more. Hence the shape of the flow outline is not scale-independent, but becomes less uneven as the measuring distance becomes larger, coinciding with a change in the chief shape-controlling factor from the geometry of crustal fragments to bulk lava rheology.

The connection between flow type and margin geometry opens the possibility of using geometric measurements to infer styles of lava emplacement, especially important for investigating inaccessible volcanoes with aerial and spaceborne techniques. However, apart from initial measurements on predominantly basaltic lavas, few data are available on the fractal dimensions (degrees of unevenness) of lava flow margins, and so the full promise of this approach must await future studies.

Forecasting the Behaviour of aa Flows

The three main types of lava surface are associated with distinct styles of flow growth. Since lava surfaces are easily monitored in the field, it would be convenient if the conditions for producing a specific surface type could also provide limits upon the likely distance and velocity a given flow might travel. Such a state of affairs does exist for aa flows, a happy circumstance since aa flows are the lava type that has most frequently threatened human activity, and they are also intermediate between pahoehoe and blocky flows. Understanding the limits to aa behaviour thus provides clues to the limiting behaviour of the other flow types.

The aa surface criterion can be used to estimate the advance velocity needed to keep breaking lava crust. The time for which a lava flow can continue to advance is controlled either by the duration of eruption (for short-lived aa flows) or, during long eruptions, by the time needed for the flow to acquire a solid front. Combining the requirement for a lava front to solidify (which gives the longest time for advance) with the aa surface criterion, it is possible to link the maximum potential length (L_m) of an aa flow to lava properties, mean underlying slope (angle β) and mean rate of discharge (Q). These expressions are:

$$L_m \approx [(1.5\,\varepsilon S / \rho\, g)^2 (\Sigma \sigma \Theta^3) / (\kappa\, \rho\, c_p)] / \sin^2 \beta$$

$$L_m \approx [3\,\varepsilon S/\rho g \kappa]^{1/2} Q^{1/2}$$

where the remaining symbols are defined in table.

In both expressions, the terms in square brackets describe the physical properties of the lava crust and are approximately constant for a given lava composition. Thus the maximum potential length is expected to increase as the mean underlying slope decreases (Equation) and as the discharge rate increases (Equation), two trends which agree well with observation.

Table: Parameters in Flow Equations.

Symbol	Meaning	Units	Nominal Value for Basalt
β	Slope angle	o	1150 (during crystallization)
c_p	Specific heat capacity	J kg^{-1} K^{-1}	
D	Thickness	m	
ε	Extension before failure	-	min. 10^{-3} (for chilled crust
εS	Energy per unit volume for failure	J m^{-3}(or Pa)	2 x 104 (during crystallization)
g	Gravitational acceleration	m s^{-2}	9.81
κ	Thermal diffusivity	m^2 s^{-1}	4.2 x 10^{-7}
L_m	Maximum potential length of flow	m	
Q	Mean discharge rate along flow	M^3 s^{-1}	
Θ	Eruption temperature (absolute)	K	1350 - 1400
ρ	Density of lava crust (not whole flow	Kg m^{-3}	2200 (~20% vol. vesicles)
S	Tensile strength	Pa	max. 10^7 (for chilled crust)
Σ	Surface emissivity	-	1
σ	Stefan-Boltzmann Constant	J m^{-2} s^{-1} K^{-4}	5.67 x 10^{-8}
t	Time	s	

The length-slope trend is counter-intuitive. Other factors being equal, it might be supposed that steeper slopes would favour longer flows. The inverse trend arises because, to keep breaking their surfaces, aa flows must maintain greater thicknesses on shallower slopes. Thicker flows need longer times for their fronts to solidify. As a result, the maximum potential length increases with decreasing slope. In practice, such lengths are usually not achieved (especially on slopes with angles smaller than 6°) because the eruption finishes or breaching occurs in the meantime. Nevertheless, the length-slope trend is important for hazard analyses because it provides realistic estimates of potential flow length when only the mean slope is known. Accordingly, it is well-suited for preparing hazard maps before an eruption occurs.

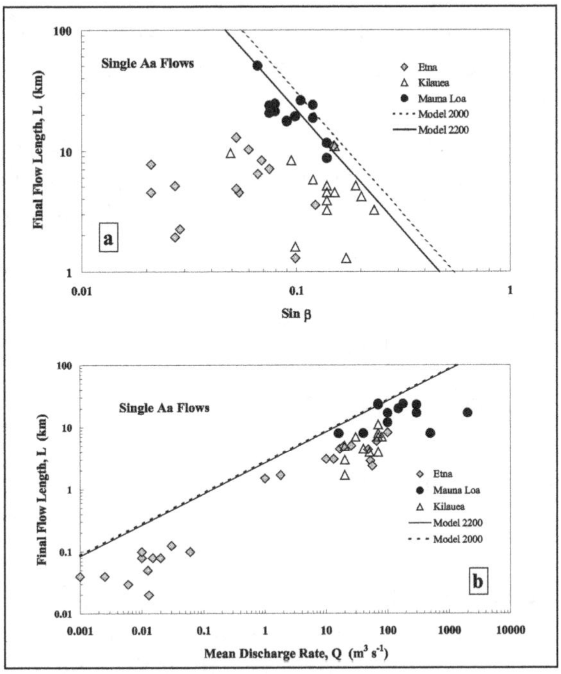

The conditions that (1) aa surfaces must break before they cool to their maximum strength and (2) flow fronts advance until solidified yield criteria for relating the maximum potential length of a single aa flow to (a) underlying slope (sinβ) and (b) rate of discharge (Q). These criteria (Equations), shown by the solid and dashed trends, agree well with observations from Etna and Hawaii. The solid and dashed lines refer to respective mean crustal densities of 2,200 and 2,000 kg $^{m-3}$.

The length-discharge rate trend may also have predictive value in some special circumstances. At several mature effusive volcanoes, the mean rate of lava effusion generally decreases with increasing altitude of the vent. Equation can therefore be modified to link maximum potential flow length to vent altitude, shorter maxima being associated with higher elevations. Pre-eruptive maps of lava hazard can thus be prepared on the basis of vent elevation, and these provide a useful check on evaluations determined from mean slopes.

Equation may also be used for short-term forecasts while an eruption is in progress. During the 1991-93 effusion on Etna, for example, secondary aa flows began to overtop an artificial barrier and descend toward Zafferana, a town almost 3 km further downslope. Initial observations suggested that it was unlikely for the mean discharge rate to exceed 1 $m^3 s^{-1}$ along any flow. From equation, a maximum flow length of about 2 km was anticipated, indicating that the immediate threat to Zafferana was small. A forecast was issued to that effect and, in the event, the longest secondary flow travelled only 1.8 km beyond the barrier.

The success of equations is remarkable not only for the simplicity of their underlying assumptions, but also because they use mean values of slope and discharge rate to determine length. In reality, surface slopes are uneven, while discharge rate changes with time and position along a flow. The agreement between theory and observation thus suggests that local variations have normally a secondary effect on maximum potential flow length, so that mean values are sufficient for most purposes. Extreme variations, such as flow over a cliff or through a very narrow ravine, may induce significant changes from the model results and, until more sophisticated analyses are available, these rare cases must be evaluated individually.

The expressions for maximum length assume that an aa flow travels until its front has solidified. The maximum time t required is:

$$t \approx [(\varepsilon S/\rho g](\rho\ c_p / \Sigma\sigma\Theta^3)^{1/2}]/\sin Q\beta.$$

As before, the combination of physical properties in the square brackets is roughly constant for a particular lava composition and, for typical angles of 2°-10° for the slopes of basaltic volcanoes, yields lengthening times of about 1-10 days (the longer times corresponding to smaller slopes). During a long-lived eruption, therefore, major breaching and the propagation of new flows is most likely to commence within days following the start of effusion. Since new flows rarely extend a flow-field by more than half the length of the first major flow, another forecasting problem is to assess the final width of a flow field, which may stretch to several kilometres.

An aa flow field can widen until halted by topography or by the end of effusion. The topographic control depends on local conditions and must be assessed on a case-by-case basis. The likely duration of an effusion can be estimated by comparison with previous eruptions, but even on well-studied volcanoes, such as Etna, Kilauea and Mauna Loa, such estimates are rarely better than inspired guesswork. Improvements will follow when it becomes possible either to estimate (probably by geophysical monitoring) the quantity of magma below ground that is available for eruption, or to forecast the long-term decay in rate of effusion as an eruption proceeds.

Forecasting the Behaviour of Pahoehoe and Blocky Flows

It is possible before an eruption to estimate the maximum length of an aa flow because the surface criterion provides a velocity, and the solidification control yields a time. Such a combination is not available either to pahoehoe or to blocky flows and so forecasts of their growth is inherently more difficult.

For pahoehoe flows, the requirement to form a continuous crust can be used to constrain maximum rates of advance to about 1 km per day on slopes typical of basaltic volcanoes.

Unfortunately, the burrowing mechanism, by which new lava intrudes into the front after travelling through insulated tubes, has so far defied realistic estimates of maximum cooling times. As it happens, observed pahoehoe flows (mostly from Hawaii) have frequently lengthened for the whole of an eruption, even when this has continued for several months. In practice, therefore, pahoehoe emplacement is often limited by the available supply of magma, and so forecasting flow behaviour is subject to the same problems encountered when estimating the final width of an aa flow field.

Among blocky lavas, surface fracturing appears to be controlled by the strength of lava beneath the chilled crust. As a result, fracturing is not required to be faster than the rate of surface chilling, and so no surface criterion is available for forecasting mean rates of advance. Velocity forecasts must therefore rely on knowing the rheological state of a lava and the rate at which it is effused, factors which can be estimated by analogy with previous effusions from the same volcano.

Since the lava is viscous and slow moving, a blocky front is expected to solidify at a rate controlled by conductive heat transfer to the breaking surface. From standard conduction theory, a crude estimate of the solidification time for common final thicknesses of 10-30 m is from six months to over four years, lengths of time that are greater than the usual durations for eruptions of blocky lava. Once again, the volume of magma available appears crucial for determining how far a lava can travel, and estimating this volume remains a fundamental obstacle to improving preeruptive forecasts of blocky flow growth.

Turbulent Lava Flows

Lava flows witnessed in historical time have had physical characteristics that favour laminar motion, and so it is this motion which has been assumed in developing quantitative models of lava behaviour. However, some ancient lavas on Earth, erupted for the most part more than 2,500 million years ago, may have been sufficiently fluid upon eruption for turbulent advance. Turbulent flow involves the spontaneous intermingling of neighbouring packets of lava, so that internal lava cools and solidifies more quickly than would have been the case for simple laminar flow. The net result may well have been the persistent surface disruption of a lava undergoing wholesale internal solidification, analogous to conditions for aa lava flow emplacement (for which, internal solidification is induced by factors other than spontaneous turbulence).

Notable candidates for ancient turbulent flows are komatiite lavas which, named after their discovery near the River Komati in South Africa, are found dotted across the world's oldest continents (Africa, Australia, and North and South America) and are often associated with economically important nickel deposits. Essentially basalts with unusually large amounts of magnesium, iron and aluminium, these lavas since their eruption have been buried by younger material and grossly deformed by major changes at the Earth's surface. As a result, it has not been possible to investigate directly the influence of turbulent motion on lava flow morphology. This is unfortunate, since interest has been generated by speculation that komatiites could be among the types of lava flow observed on other planets. Only when direct chemical analyses are available for extraterrestrial flows will it be possible to gauge the potential signifance of turbulent flow on volcanoes beyond Earth.

Lava Flows and Planetary Development

Lava flow fields are among the commonest surface features of the terrestrial planets. Quite apart from hazard analyses, therefore, an understanding of flow emplacement is essential for investigating the evolution of planetary surfaces. By virtue of their remote location, most information on extraterrestrial flows concerns their dimensions and geometry. Especially impressive is the fact that flows stretching hundreds of kilometres are commonplace on the Moon, Mars and Venus, some exceeding the size of ancient flood basalts on Earth. Primary goals of planetary studies are to use flow shape and size to deduce chemical composition and to infer conditions in the magmatic feeding systems, thereby probing also into the states of planetary crusts and mantles.

Despite their large size, the shapes of planetary flow fields resemble those of their smaller cousins on Earth. Since flow shape depends on the ratios of controlling factors, geometrical similarity suggests that these factors remain within the same range of proportions to each other as they occur on Earth. It is thus realistic to assume that, unless extraterrestrial lava flows have been emplaced exclusively under turbulent conditions, the range of controlling ratios (e.g., among the forces driving and resisting motion) found on Earth can be applied to the neighbouring planets and, hence, that extraterrestrial flows will also belong to the pahoehoe-aa-blocky spectrum.

Lava Tubes

Lava tubes are self-forming within a flow field, and like channels they develop downflow during an eruption. Lava tubes are very efficient transporters of lava from the vent to the flow front, and Laszlo Keszthelyi has recently calculated that lava flowing in a tube loses only about 1C° of heat/kilometer. At the flow front, the lava behaves much like a river delta, forming small distributary tubes that continue to branch until they consist of the same type of single flow-unit tubes (toes) that have been forming the flow the whole way downslope.

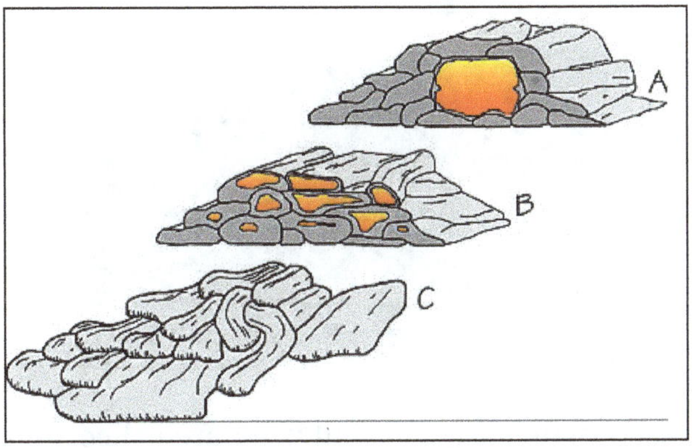

Above: Diagrammatic cross-sections through a pahoehoe flow (same time, different locations; grey is solid, reds are still fluid). A: Master tube formed within a pile of flow units, B: distributary tube system near flow front, continued flow in some of these will result in their merging to form a master tube, C: flow front consisting of numerous pahoehoe toes.

The amount of gas in lava at the flow front is somewhat lower than that at the vent because although lava tubes are good thermal insulators, they are not so well sealed that gas can be trapped. Gas escapes through cracks in the tube and also through skylights, places where the roof of a tube has fallen in to expose the flowing lava. Hot escaping fumes are strongly concentrated at skylights and they must be approached very carefully.

An active skylight formed on the main lava tube from the Kupa'ianaha vent, Kilauea east rift zone, November 1991. The skylight is about 2 m across.

Whenever the flow front of a flow stagnates on sub-horizontal surfaces (such as on the coastal plain), it starts to grow by inflation as well as by spreading. This growth by inflation has only recently been recognized as an important factor in flow field growth. The initial advancing flow front is usually <1 m thick. Subsequent inflation of the flow can increase its thickness up to ~10 m.

Places where inflation is particularly concentrated are called tumuli (singular: tumulus). Tumuli range in size from a few square meters in area and ~1 m high to a few thousand square meters in area and ~10 m high. Tumuli are often ubiquitous on flat-lying portions of a pahoehoe flow field. There are also places that inflate less than the surrounding area. These are called lava-rise pits. A large tumulus developed in the 1859 Mauna Loa flow near the coast.

The lava of an advancing pahoehoe flow front is relatively vesicle-rich (although less gas-rich than near-vent "shelly" pahoehoe).

Shelly pahoehoe: Note that the surface skin is only a few cm thick. The lava has not drained away from under the skin but rather within a rapidly quenched outer skin, an initially foam-like mixture of lava and gas bubbles separated into a gas-only top and lava-only bottom. It is easy to see why walking on shelly pahoehoe is not a pleasant pastime.

S-type pahoehoe flow unit. The scale indicates 13 cm.

This bubble-rich flow front pahoehoe has recently been termed s-type pahoehoe, the "s" standing for "spongy" which aptly describes its appearance in cross-section. While the lava is still molten, vesicles often coalesce just beneath the quenched skin and this forms a mostly gas (and therefore weak) layer under the top 1-2mm of surface glass. This weak surface layer of s-type pahoehoe spalls off, sometimes only within a few days, exposing the texture of the underlying vesicles.

The vesicularity is also evident on the top surface of the lava, which consists of filamentous strands of vesicle walls that have been stretched as the skin deforms.

Close-up of the surface of an s-type pahoehoe toe showing the filamentous structure that forms from stretched an deformed vesicle walls.

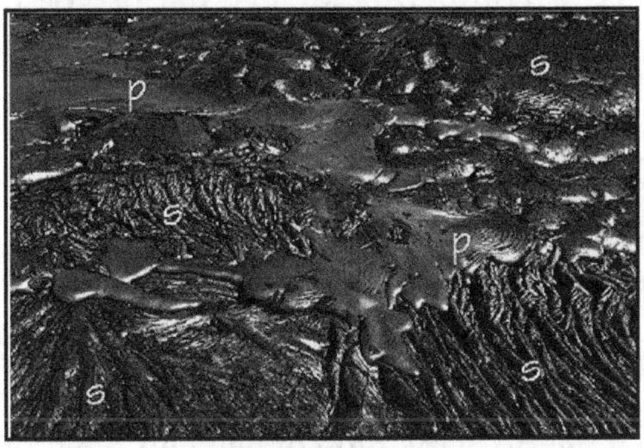

The lava that is stored within a flow field before being erupted, on the other hand, is distinctly vesicle-poor. It has been termed p-type pahoehoe because the bottoms of these flow units commonly contain pipe vesicles. Flow units of p-type pahoehoe have an appearance that makes them appear to be less viscous and their thickness is often <10 cm. This low-lying appearance is mainly due to the small number of gas bubbles within the p-type pahoehoe; if you stir it with a rock hammer it definitely "feels" more viscous than the more common gas-rich s-type pahoehoe.

S-type and p-type pahoehoe together. Note the thinness of the p-type lava flow units compared to the more billowy and/or ropy forms of the s-type. The blue color is distinctive of fresh p-type pahoehoe.

When first erupted, p-type pahoehoe has a distinct shiny blue color. During weathering it develops a patchy ochre coating.

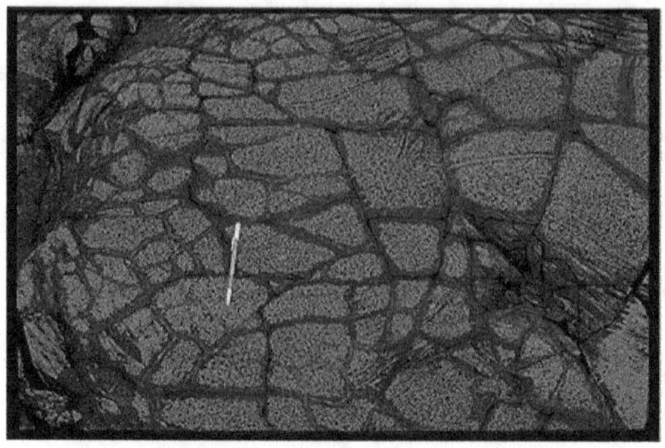

Mauna Iki p-type pahoehoe erupted in 1920, showing the yellowish secondary mineral that usually develops on this type of lava. Note that the secondary mineral does not form near fractures, perhaps because it dissolves in rainwater that accumulates there.

The top layer of this vesicle-poor lava strongly resembles obsidian and was originally thought to form when heavy rain falls while the lava is flowing. Because you can see p-type pahoehoe flowing during any weather, this idea can easily be shown to be false. The most commonly held idea is that the paucity of vesicles is due to the prolonged storage within the flow field (up to a few days) prior to being forced onto the surface by an influx of new lava. During this storage gas bubbles are able to migrate upward and escape through fractures in the surface carapace. A competing idea is that p-type pahoehoe is actually richer in gas; the overburden of the solid flow surface has caused the gas to go back into solution.

Volcanic Ash

Volcanic ash is the ash produced when an eruption is about to or is taking place. There are several negative effects of volcanic ash on the people that live in the area and even those that are nowhere near the volcano. This type of ash can have problems as far as different continents with several instances in which it has actually dropped the overall temperature of the planet. The amount of problems that the ash will cause depends greatly on the size of the eruption but even a smaller eruption could have measurable effects on an area.

Real Danger

When a lot of people hear about volcanic ash they may think that it will be mostly a problem that has to do with cleanup. In fact, the problems could be a lot worse. In some cases the problems that can arise from the ash could be deadly. In some instances when ash has fallen it has destroyed plant and animal life. That is because plant life can get buried under volcanic ash while the animals that depend on that plant life will starve to death. Another problem could be that the animals in the area will be inhaling the ash which can interfere with their respiratory system, not allowing them to breathe properly and therefore killing them.

Another problem which volcanic ash can create in an area is when it falls on the roofs of the people who live in the area. The problem with that would be the fact that it can weigh down on weaker structures and therefore cause them to collapse under their own weight. The problem can get even worse if rain were to fall at the same location, because the rain would solidify the ash and therefore make it heavier. Finally the people in the surrounding area could suffer the same fate as the animal life. If they don't take good care of themselves and breathe in the ashes, those ashes could solidify inside their lungs, making it a potentially deadly situation.

Pyroclastic Flow

One of the very real dangers of volcanic ash is the possibility of pyroclastic flow. That would be a mass of gas, ash and rock that could travel at speeds close to 125 miles per hour. The problem with this type of situation is not only the speed of the gas and the magnitude but also the temperature of it. It has been calculated that the temperature inside a pyroclastic flow is anywhere between 752 to 1,472 degrees Fahrenheit. That means that anything in its path would cook almost immediately. Signs of this type of event were found by archaeologists who were studying the eruption of Mount Vesuvius in 79 A.D.

Effects on Airplanes

Airplanes can be greatly affected by volcanic ash, especially because engines and ash do not mix. If ash goes into a plane's engine it can stop the engines and cause the aircraft to crash. When the ash solidifies in the engine it will create crystals that will damage the engine and that is what stops it all together. Pilots who have flown into volcanic ash have reported the plane's throttle system fails and as they apply more throttle the situation tends to get even worse because that is what cools down the ash, solidifying it.

Impact on Everyday Life

Volcanic ash can be as dangerous as to cause a drop in worldwide temperatures. If the eruption is big enough then that drop in temperature could impact crops and communications and cause food shortages.

Volcanoes Hotspots

In geology, the places known as hotspots or hot spots are volcanic regions thought to be fed by underlying mantle that is anomalously hot compared with the surrounding mantle. They may be on, near to, or far from tectonic plate boundaries. Currently, there are two hypotheses that attempt to explain their origins. One suggests that they are due to hot mantle plumes that rise as thermal diapirs from the core–mantle boundary. An alternative hypothesis postulates that it is not high temperature that causes the volcanism, but lithospheric extension that permits the passive rising of melt from shallow depths. This hypothesis considers the term "hotspot" to be a misnomer, asserting that the mantle source beneath them is, in fact, not anomalously hot at all. Well known examples include Hawaii and Yellowstone.

The origins of the concept of hotspots lie in the work of J. Tuzo Wilson, who postulated in 1963 that the Hawaiian Islands result from the slow movement of a tectonic plate across a hot region beneath the surface. It was later postulated that hotspots are fed by narrow streams of hot mantle rising from the Earth's core–mantle boundary in a structure called a mantle plume. Whether or not such mantle plumes exist is currently the subject of a major controversy in Earth science. Estimates for the number of hotspots postulated to be fed by mantle plumes has ranged from about 20 to several thousands, over the years, with most geologists considering a few tens to exist. Hawaii, Réunion, Yellowstone, Galápagos, and Iceland are some of the currently most active volcanic regions to which the hypothesis is applied.

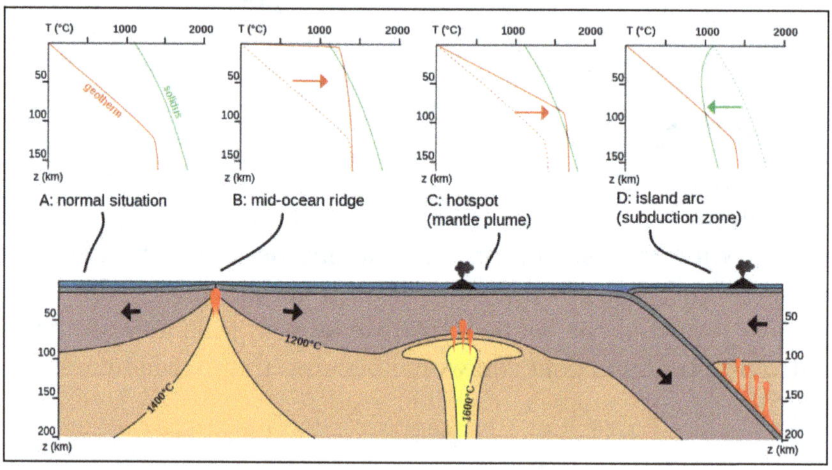

Schematic diagram showing the physical processes inside the Earth that lead to the generation of magma. Partial melting begins above the fusion point.

Most hotspot volcanoes are basaltic (e.g., Hawaii, Tahiti). As a result, they are less explosive than subduction zone volcanoes, in which water is trapped under the overriding plate. Where hotspots occur in continental regions, basaltic magma rises through the continental crust, which melts to form rhyolites. These rhyolites can form violent eruptions. For example, the Yellowstone Caldera was formed by some of the most powerful volcanic explosions in geologic history. However, when the rhyolite is completely erupted, it may be followed by eruptions of basaltic magma rising through the same lithospheric fissures (cracks in the lithosphere). An example of this activity is the Ilgachuz Range in British Columbia, which was created by an early complex series of trachyte and rhyolite eruptions, and late extrusion of a sequence of basaltic lava flows. The hotspot hypothesis is now closely linked to the mantle plume hypothesis.

Comparison with Island Arc Volcanoes

Hotspot volcanoes are considered to have a fundamentally different origin from island arc volcanoes. The latter form over subduction zones, at converging plate boundaries. When one oceanic plate meets another, the denser plate is forced downward into a deep ocean trench. This plate, as it is subducted, releases water into the base of the over-riding plate, and this water mixes with the rock, thus changing its composition causing some rock to melt and rise. It is this that fuels a chain of volcanoes, such as the Aleutian Islands, near Alaska.

Hotspot Volcanic Chains

The joint mantle plume/hotspot hypothesis envisages the feeder structures to be fixed relative to one another, with the continents and seafloor drifting overhead. The hypothesis thus predicts that time-progressive chains of volcanoes are developed on the surface. Examples are Yellowstone, which lies at the end of a chain of extinct calderas, which become progressively older to the west. Another example is the Hawaiian archipelago, where islands become progressively older and more deeply eroded to the northwest.

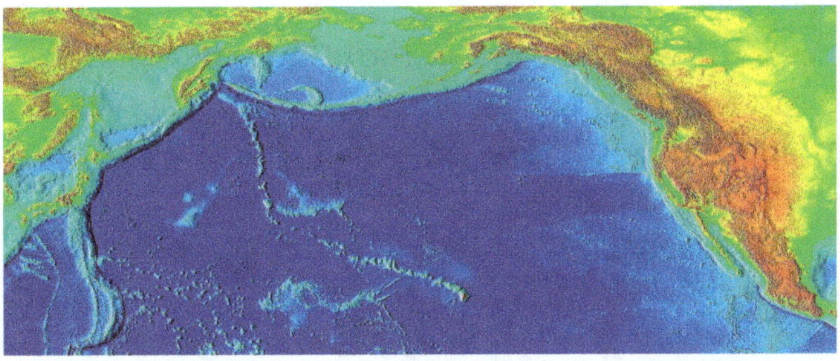

Over millions of years, the Pacific Plate has moved over the Hawaii hotspot, creating a trail of underwater mountains that stretch across the Pacific.

Geologists have tried to use hotspot volcanic chains to track the movement of the Earth's tectonic plates. This effort has been vexed by the lack of very long chains, by the fact that many are not time-progressive (e.g. the Galápagos) and by the fact that hotspots do not appear to be fixed relative to one another (e.g. Hawaii and Iceland).

References

- Volcano, science: britannica.com, Retrieved 29 July, 2019
- Kilburn-Lavas-Encyc-Volcanoes-2000, volcanoscope: ucl.ac.uk, Retrieved 30 August, 2019
- Lava-tubes: volcano.oregonstate.edu, Retrieved 31 January, 2019
- Volcanic-ash: basicplanet.com, Retrieved 1 February, 2019
- Reading-volcanoes-hotspots, chapter, geo: courses.lumenlearning.com, Retrieved 2 March, 2019

2
Types of Volcanoes

Some of the major types of volcanoes are submarine volcanoes, strato volcanoes, shield volcanoes, cryovolcanoes, ice volcanoes, mud volcanoes, supervolcanoes and subglacial volcanoes. This chapter closely examines these types of volcanoes to provide an extensive understanding of the subject.

Submarine Volcanoes

A submarine volcano is different than all of the other types because they are found underwater. These volcanoes are in fact some of the most productive of the volcanic systems, producing around three quarters of the year's output of magma but most people do not think of them because they are usually located around 8,500 feet underwater on average. Some scientists estimate that in the Pacific Ocean in just a million square kilometers of the ocean floor you will find around 4,000 volcanoes. Like other types, a submarine volcano can vary greatly in height, although it is estimated that around 75,000 of those in the Pacific Ocean rise half a mile over the ocean floor.

Eruptions

Although each submarine volcano is located underwater, it will still produce eruptions while active and the products of these eruptions are what shape the way the sea floor looks. Scientists estimate that out of the million or so of these volcanoes that can be found around the world, many thousands of them are active. Despite this because of their location underwater, it can be hard to catch a submarine volcano in the eruption process. However, scientists do know that most eruptions from these volcanoes will take place in shallow water although recent technological advances have helped scientists document those that take place in deeper water well.

Deposits

Because a submarine volcano is located underwater, the deposits that are created during the eruption can be very different than those of other types of volcanoes. Volcanologists are actually unsure if an explosive eruption of a submarine volcano is even possible. That is because due to the high pressure created by the water (250 times the atmospheric pressure) bubbles are hard to form in the magma and lava. Because the water will almost instantly quench the lava, the outer surface of volcanic flows becomes glass.

Lava Flows

The shape of the flows of a submarine volcano tends to be pillow-like as compared to the blocky flows that are typical of land volcanoes. These pillows are created when the crust of a flow will split causing the lava to ooze out at which point it flows a very short distance and solidifies. The process then repeats with the new piece of lava and ends up creating what seems to be a large stack of pillows. The lava flows can also take other shapes, such as sheet flows and tumuli, which is an inflation feature.

Volcanic Ecosystems

Like all volcanoes, a submarine volcano is associated with a unique volcanic ecosystem. These are usually located by hydrothermal vents, or black smokers, which have been located at all of the mid-ocean ridges around the world. At these vents, hydrogen sulfide and water in addition to other minerals exit these vents, creating an ecosystem of organisms that live using sulfur instead of sunlight. Here you can find tube worms, mussels, giant clams and other organisms.

Detection of Eruptions

Despite recent advances in technology it is still difficult for scientists to detect the eruption of a submarine volcano. In fact, they have only been capable of detecting them at all since around the year 1990. Long before they were first detected, oceanographers and volcanologists alike were aware that submarine volcano eruptions were occurring but had no way to know where and when. In recent decades, however, they created a new system that measures the small earthquakes that are normally part of magma migration.

Instead of relying on feeling vibrations like they do with earthquakes on land, this system listens for them as the vibrations underwater will produce an acoustic wave. Since this system was created, it has been much easier for scientists to pinpoint the location of these underwater eruptions.

Strato Volcanoes

A stratovolcano, also known as a composite volcano, is a conical volcano built up by many layers (strata) of hardened lava, tephra, pumice and ash. Unlike shield volcanoes, stratovolcanoes are characterized by a steep profile with a summit crater and periodic intervals of explosive eruptions and effusive eruptions, although some have collapsed summit craters called calderas. The lava flowing from stratovolcanoes typically cools and hardens before spreading far, due to high viscosity. The magma forming this lava is often felsic, having high-to-intermediate levels of silica (as in rhyolite, dacite, or andesite), with lesser amounts of less-viscous mafic magma. Extensive felsic lava flows are uncommon, but have travelled as far as 15 km (9.3 mi).

Stratovolcanoes are sometimes called "composite volcanoes" because of their composite stratified structure built up from sequential outpourings of erupted materials. They are among the most common types of volcanoes, in contrast to the less common shield volcanoes. Two famous examples of stratovolcanoes are Krakatoa in Indonesia, known for its catastrophic eruption in 1883 and Vesuvius in Italy, whose catastrophic eruption in AD 79 ruined the Roman cities of Pompeii and Herculaneum. Both eruptions claimed thousands of lives. In modern times, Mount Saint Helens and Mount Pinatubo have erupted catastrophically, with fewer deaths.

The possible existence of stratovolcanoes on other terrestrial bodies of the Solar System has not been conclusively demonstrated. The one feasible exception is the existence of some isolated massifs on Mars, for example the Zephyria Tholus.

Creation

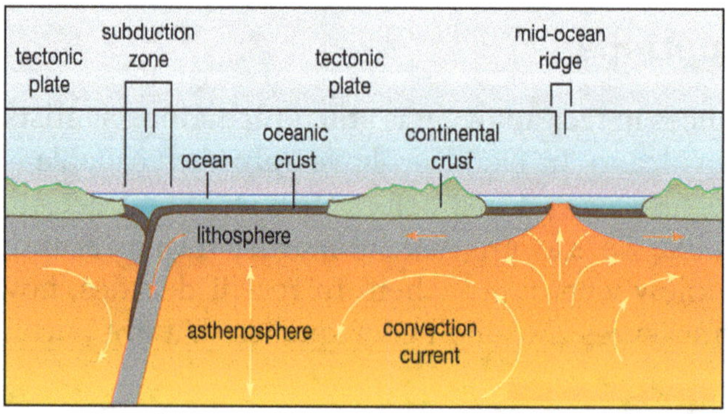

Cross-section of subduction zone and associated stratovolcanoes.

Stratovolcanoes are common at subduction zones, forming chains and clusters along plate tectonic boundaries where oceanic crust is drawn under continental crust (continental arc volcanism, e.g. Cascade Range, Andes, Campania) or another oceanic plate (island arc volcanism, e.g. Japan, Philippines, Aleutian Islands). The magma forming stratovolcanoes rises when water trapped both in hydrated minerals and in the porous basalt rock of the upper oceanic crust is released into mantle rock of the asthenosphere above the sinking oceanic slab. The release of water from hydrated minerals is termed "dewatering", and occurs at specific pressures and temperatures for each mineral, as the plate descends to greater depths. The water freed from the rock lowers the melting point of the overlying mantle rock, which then undergoes partial melting and rises due to its lighter density relative to the surrounding mantle rock, and pools temporarily at the base of the lithosphere. The magma then rises through the crust, incorporating silica-rich crustal rock, leading to a final intermediate composition. When the magma nears the top surface, it pools in a magma chamber within the crust below the stratovolcano.

There, the relatively low pressure allows water and other volatiles (mainly CO_2, SO_2, Cl_2, and H_2O) dissolved in the magma to escape from solution, as occurs when a bottle of carbonated water is opened, releasing CO_2. Once a critical volume of magma and gas accumulates, the plug (solidified blockage) of the volcanic vent is broken, leading to a sudden explosive eruption.

Hazards

Mount Etna on the island of Sicily, in southern Italy.

In records, explosive eruptions at subduction zone (convergent-boundary) volcanoes have posed the greatest hazard to civilizations. Subduction-zone stratovolcanoes, such as Mount Saint Helens, Mount Etna and Mount Pinatubo, typically erupt with explosive force: the magma is too stiff to allow easy escape of volcanic gases. As a consequence, the tremendous internal pressures of the trapped volcanic gases remain and intermingle in the pasty magma. Following the breaching of the vent and the opening of the crater, the magma degasses explosively. The magma and gases blast out with high speed and full force.

Since 1600 CE, nearly 300,000 people have been killed by volcanic eruptions. Most deaths were caused by pyroclastic flows and lahars, deadly hazards that often accompany explosive eruptions of subduction-zone stratovolcanoes. Pyroclastic flows are swift, avalanche-like, ground-sweeping, incandescent mixtures of hot volcanic debris, fine ash, fragmented lava and superheated gases that can travel at speeds in excess of 160 km/h (100 mph). Around 30,000 people were killed by pyroclastic flows during the 1902 eruption of Mount Pelée on the island of Martinique in the Caribbean. In March to April 1982, three explosive eruptions of El Chichón in the State of Chiapas in southeastern Mexico, caused the worst volcanic disaster in that country's history. Villages within 8 km (5 mi) of the volcano were destroyed by pyroclastic flows, killing more than 2,000 people.

Mount Fuji on Honshu (left) and Mount Unzen on Kyushu (right), two of Japan's stratovolcanoes.

Two Decade Volcanoes that erupted in 1991 provide examples of stratovolcano hazards. On June 15, Mount Pinatubo spewed an ash cloud 40 km (25 mi) into the air and produced huge pyroclastic surges and lahar floods that devastated a large area around the volcano. Pinatubo, located in Central Luzon just 90 km (56 mi) west-northwest from Manila, had been dormant for 6 centuries before the 1991 eruption, which ranks as one of the largest eruptions in the 20th century. Also in 1991, Japan's Unzen Volcano, located on the island of Kyushu about 40 km (25 mi) east of Nagasaki, awakened from its 200-year slumber to produce a new lava dome at its summit. Beginning in June, repeated collapse of this erupting dome generated ash flows that swept down the mountain's slopes at speeds as high as 200 km/h (120 mph). Unzen is one of more than 75 active volcanoes in Japan; an eruption in 1792 killed more than 15,000 people—the worst volcanic disaster in the nation's history.

The eruption of Mount Vesuvius in 79 completely smothered the nearby ancient cities of Pompeii and Herculaneum with thick deposits of pyroclastic surges and lava flows. Although death toll is estimated between 13,000 and 26,000 remains, the exact number is still unclear. Vesuvius is recognized as one of the most dangerous volcanoes, due to its capacity for powerful explosive eruptions combined with the high population density of the surrounding Metropolitan Naples area (totaling about 3.6 million inhabitants).

Ash

Snow-like blanket of Mount Pinatubo's ashfall deposits.

Apart from possibly affecting the climate, volcanic clouds from explosive eruptions also pose a serious hazard to aviation safety. For example, during the 1982 eruption of Galunggung in Java, British Airways Flight 9 flew into the ash cloud, suffering temporary engine failure and structural damage. During the past two decades, more than 60 airplanes, mostly commercial airliners, have been damaged by in-flight encounters with volcanic ash. Some of these encounters have resulted in the power loss of all engines, necessitating emergency landings. Luckily, to date no crashes have happened because of jet aircraft flying into volcanic ash. Ashfalls are a threat to health when inhaled and is also a threat to property with enough accumulation. An accumulation of 30 cm (12 in) is sufficient to cause most buildings to collapse. Dense clouds of hot volcanic ash, caused by the collapse of an eruptive column or by being laterally expelled from the partial collapse of a volcanic edifice or lava dome during explosive eruptions, can generated devastating pyroclastic flows or surges, which can sweep off everything in their paths.

Lava

Mayon Volcano extruding lava flows during its eruption.

Lava flows from stratovolcanoes are generally not a significant threat to humans and animals because the highly viscous lava moves slowly enough for everyone to flee out of the path of flow. The lava flows are more of a threat to property. However, not all stratovolcanoes erupt viscous and sticky lava. Nyiragongo is very dangerous because its magma has an unusually low silica content, making it quite fluid. Fluid lavas are typically associated with the formation of broad shield volcanoes such as those of Hawaii, but Nyiragongo has very steep slopes down which lava can flow at up to 100 km/h (60 mph). Lava flows could melt down ice and glaciers that accumulated on the volcano's crater and upper slopes, generating massive lahar flows. Rarely, generally fluid lava could also generate massive lava fountains, while lava of thicker viscosity can solidify within the vent, creating a block which can result in highly explosive eruptions.

Volcanic Bombs

Volcanic bombs are extrusive igneous rocks ranging from the size of books to small cars, that are explosively ejected from stratovolcanoes during their climactic eruptive phases. These "bombs" can travel over 20 km (12 mi) away from the volcano, and present a risk to buildings and living beings while shooting at very high speeds (hundreds of kilometers/ miles per hour) through the air. Most bombs do not themselves explode on impact, but rather carry enough force so as to have destructive effects as if they exploded.

Lahar

Lahars (from a Javanese term for volcanic mudflows) are mixtures of volcanic debris and water. Lahars usually come from two sources: rainfall or the melting of snow and ice by hot volcanic elements, such as lava. Depending on the proportion and temperature of water to volcanic material, lahars can range from thick, gooey flows that have the consistency of wet concrete to fast-flowing, soupy floods. As lahars flood down the steep sides of stratovolcanoes, they have the strength and speed to flatten or drown everything in their paths. Hot ash clouds, lava flows and pyroclastic surges ejected during 1985 eruption of Nevado del Ruiz in Colombia melted snow and ice atop the 5,321 m (17,457 ft) high Andean volcano. The ensuing lahar flooded the city of Armero and nearby settlements, killing 25,000 people.

Effects on Climate and Atmosphere

Paluweh eruption.

As per the examples, while the Unzen eruptions have caused deaths and considerable local damage in the historic past, the impact of the June 1991 eruption of Mount Pinatubo was global. Slightly cooler-than-usual temperatures were recorded worldwide, with brilliant sunsets and intense sunrises attributed to the particulates; this eruption lofted particles high into the stratosphere. The aerosols that formed from the sulfur dioxide (SO_2), carbon dioxide (CO_2), and other gases dispersed around the world. The SO_2 mass in this cloud—about 22 million tons—combined with water (both of volcanic and atmospheric origin) formed droplets of sulfuric acid, blocking a portion of the sunlight from reaching the troposphere and ground. The cooling in some regions is thought to have been as much as 0.5 °C (0.9 °F). An eruption the size of Mount Pinatubo tends to affect the weather for a few years; the material injected into the stratosphere gradually drops into the troposphere, where it is washed away by rain and cloud precipitation.

A similar, but extraordinarily more powerful phenomenon occurred in the cataclysmic April 1815 eruption of Mount Tambora on Sumbawa island in Indonesia. The Mount

Tambora eruption is recognized as the most powerful eruption in recorded history. Its eruption cloud lowered global temperatures by as much as 3.5 °C (6.3 °F). In the year following the eruption, most of the Northern Hemisphere experienced sharply cooler temperatures during the summer. In parts of Europe, Asia, Africa and North America, 1816 was known as the "Year Without a Summer", which caused considerable agricultural crisis and a brief but bitter famine, which generated a series of distresses across much of the affected continents.

Shield Volcanoes

A shield volcano is a type of volcano usually composed almost entirely of fluid lava flows. It is named for its low profile, resembling a warrior's shield lying on the ground. This is caused by the highly fluid (low viscosity) lava erupted, which travels farther than lava erupted from a stratovolcano, and results in the steady accumulation of broad sheets of lava, building up the shield volcano's distinctive form.

Geology

Structure

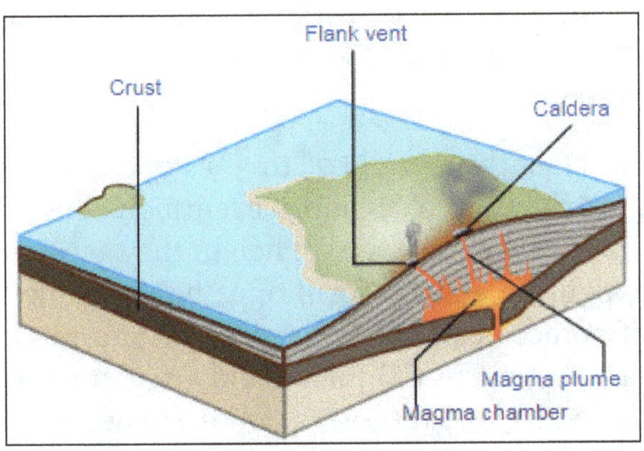

Diagram of the common structural features of a shield volcano.

Shield volcanoes are distinguished from the three other major volcanic archetypes—stratovolcanoes, lava domes, and cinder cones—by their structural form, a consequence of their unique magmatic composition. Of these four forms shield volcanoes erupt the least viscous lavas: whereas stratovolcanoes and especially lava domes are the product of highly immotile flows and cinder cones are constructed by explosively eruptive tephra, shield volcanoes are the product of gentle effusive eruptions of highly fluid lavas that produce, over time, a broad, gently sloped eponymous "shield". Although the term is generally ascribed to basaltic shields it has also at times been appended to rarer scutiform volcanoes of differing magmatic composition—principally pyroclastic shields, formed by the accumulation of fragmental material from particularly powerful explosive eruptions, and rarer felsic lava shields formed by unusually fluid felsic magmas. Examples of pyroclastic shields include Billy Mitchell volcano in Papua New Guinea and the Purico complex in Chile; an example of a felsic shield is the Big Obsidian Flow in Oregon. Shield volcanoes are also related in origination to vast lava plateaus and flood basalts present in various parts of the world, generalized eruptive features which occur along linear fissure vents and are distinguished from shield volcanoes proper by the lack of an identifiable primary eruptive center.

Active shield volcanoes experience near-continuous eruptive activity over extremely long periods of time, resulting in the gradual build-up of edifices that can reach extremely large dimensions. With the exclusion of flood basalts, mature shields are the largest volcanic features on Earth: the summit of the largest subaerial volcano in the world, Mauna Loa, lies 4,169 m (13,678 ft) above sea level, and the volcano, over 60 mi (100 km) wide at its base, is estimated to contain about 80,000 km³ (19,000 cu mi) of basalt. The mass of the volcano is so great that it has slumped the crust beneath it a further 8 km (5 mi); accounting for this subsidence and for the height of the volcano above the sea floor, the "true" height of Mauna Loa from the start of its eruptive history is about 17,170 m (56,000 ft). Mount Everest, by comparison, is 8,848 m (29,029 ft) in height. In September 2013 a team led by the University of Houston's William Sager announced the singular origination of Tamu Massif, an enormous extinct submarine shield volcano of previously unknown origin which, approximately 450 by 650 km (280 by 400 mi) in area, dwarfs all previously known volcanoes on the planet. The research has not yet been confirmed.

Shield volcanoes feature a gentle (usually 2° to 3°) slope that gradually steepens with elevation (reaching approximately 10°) before eventually flattening near the summit, forming an overall upwardly convex shape. In height they are typically about one twentieth their width. Although the general form of a "typical" shield volcano varies little worldwide regional differences exist in their size and morphological characteristics. Typical shield volcanoes present in California and Oregon measure 3 to 4 mi (5 to 6 km) in diameter and 1,500 to 2,000 ft (500 to 600 m) in height; shield volcanoes in the central Mexican Michoacán–Guanajuato volcanic field, by comparison, average 340 m (1,100 ft) in height and 4,100 m (13,500 ft) in width, with an average slope angle of 9.4° and an average volume of 1.7 km³ (0.4 cu mi).

Rift zones are a prevalent feature on shield volcanoes that is rare on other volcanic types. The large, decentralized shape of Hawaiian volcanoes as compared to their smaller, symmetrical Icelandic cousins can be attributed to rift eruptions. Fissure venting is common in Hawai'i; most Hawaiian eruptions begin with a so-called "wall of fire" along a major fissure line before centralizing to a small number of points. This accounts for their asymmetrical shape, whereas Icelandic volcanoes follow a pattern of central eruptions dominated by summit calderas, causing the lava to be more evenly distributed or symmetrical.

Eruptive Characteristics

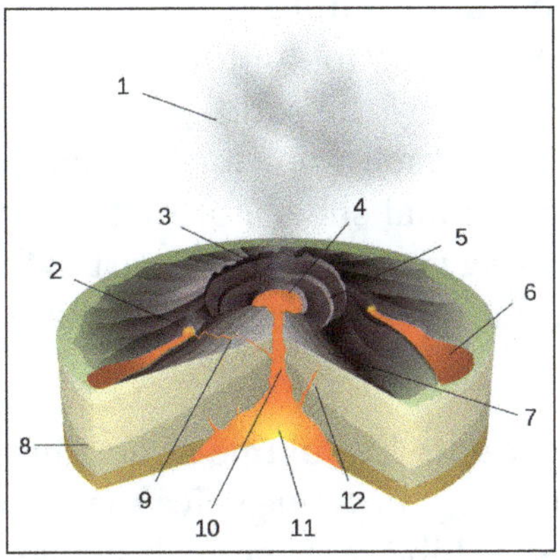

Diagram of a Hawaiian eruption: 1. Ash plume 2. Lava fountain 3. Crater 4. Lava lake 5. Fumaroles 6. Lava flow 7. Layers of lava and ash 8. Stratum 9. Sill 10. Magma conduit 11. Magma chamber 12. Dike.

Most of what is currently known about shield volcanic eruptive character has been gleaned from studies done on the volcanoes of Hawai'i island, by far the most intensively studied of all shields due to their scientific accessibility; the island lends its name to the slow-moving, effusive eruptions typical of shield volcanism, known as Hawaiian eruptions. These eruptions, the calmest of volcanic events, are characterized by the effusive emission of highly fluid basaltic lavas with low gaseous content. These lavas travel a far greater distance than those of other eruptive types before solidifying, forming extremely wide but relatively thin magmatic sheets often less than 1 m (3 ft) thick. Low volumes of such lavas layered over long periods of time are what slowly constructs the characteristically low, broad profile of a mature shield volcano.

Also unlike other eruptive types, Hawaiian eruptions often occur at decentralized fissure vents, beginning with large "curtains of fire" that quickly die down and concentrate at specific locations on the volcano's rift zones. Central-vent eruptions, meanwhile, often take the form of large lava fountains (both continuous and sporadic), which can reach heights of hundreds of meters or more. The particles from lava fountains usually cool in the air before hitting the ground, resulting in the accumulation of cindery scoria fragments; however, when the air is especially thick with clasts, they cannot cool off fast enough due to the surrounding heat, and hit the ground still hot, accumulating into spatter cones. If eruptive rates are high enough, they may even form splatter-fed lava flows. Hawaiian eruptions are often extremely long-lived; Pu'u 'Ō'ō, a cinder cone of Kīlauea, erupted continuously from January 3, 1983 until April 2018.

Flows from Hawaiian eruptions can be divided into two types by their structural characteristics: pāhoehoe lava which is relatively smooth and flows with a ropey texture, and 'a' ā flows which are denser, more viscous (and thus slower moving) and blockier. These lava flows can be anywhere between 2 and 20 m (10 and 70 ft) thick. 'A'a lava flows move through pressure— the partially solidified front of the flow steepens due to the mass of flowing lava behind it until it breaks off, after which the general mass behind it moves forward. Though the top of the flow quickly cools down, the molten underbelly of the flow is buffered by the solidifying rock above it, and by this mechanism, ʻaʻa flows can sustain movement for long periods of time. Pāhoehoe flows, in contrast, move in more conventional sheets, or by the advancement of lava "toes" in snaking lava columns. Increasing viscosity on the part of the lava or shear stress on the part of local topography can morph a pāhoehoe flow into an a'a one, but the reverse never occurs.

Although most shield volcanoes are by volume almost entirely Hawaiian and basaltic in origin, they are rarely exclusively so. Some volcanoes, like Mount Wrangell in Alaska and Cofre de Perote in Mexico, exhibit large enough swings in their historical magmatic eruptive characteristics to cast strict categorical assignment in doubt; one geological study of de Perote went so far as to suggest the term "compound shield-like volcano" instead. Most mature shield volcanoes have multiple cinder cones on their flanks, the results of tephra ejections common during incessant activity and markers of currently and formerly active sites on the volcano. One prominent such parasitic cones is Puʻu ʻŌʻō on Kīlauea—continuous activity ongoing since 1983 has built up a 2,290 ft (698 m) tall cone at the site of one of the longest-lasting rift eruptions in known history.

ʻAʻa advances over solidified pāhoehoe on Kīlauea, Hawaii.

The Hawaiian shield volcanoes and the Galápagos islands are unique in that they are not located near any plate boundaries; instead, the two chains are fed by the movement of oceanic plates over an upwelling of magma known as a hotspot. Over millions of years, the tectonic movement that moves continents also creates long volcanic trails across the seafloor. The Hawaiian and Galápagos shields, and other hotspot shields like them, are both constructed of oceanic island basalt. Their lavas are characterized by high levels of sodium, potassium, and aluminium.

Features common in shield volcanism include lava tubes. Lava tubes are cave-like volcanic straights formed by the hardening of overlaying lava. These structures help further the propagation of lava, as the walls of the tube insulates the lava within. Lava tubes can account for a large portion of shield volcano activity; for example, an estimated 58% of the lava forming Kīlauea comes from lava tubes.

A pāhoehoe lava fountain on Kīlauea erupts.

A lava lake in the caldera of Erta Ale, an active shield volcano in Ethiopia.

In some shield volcano eruptions, basaltic lava pours out of a long fissure instead of a central vent, and shrouds the countryside with a long band of volcanic material in the form of a broad plateau. Plateaus of this type exist in Iceland, Washington, Oregon, and Idaho; the most prominent ones are situated along the Snake River in Idaho and the Columbia River in Washington and Oregon, where they have been measured to be over 1 mi (2 km) in thickness.

Calderas are a common feature on shield volcanoes. They are formed and reformed over the volcano's lifespan. Long eruptive periods form cinder cones, which then collapse over time to form calderas. The calderas are often filled up by future eruptions, or formed elsewhere, and this cycle of collapse and regeneration takes place throughout the volcano's lifespan.

Interactions between water and lava at shield volcanoes can cause some eruptions to become hydrovolcanic. These explosive eruptions are drastically different from the usual shield volcanic activity, and are especially prevalent at the waterbound volcanoes of the Hawaiian Isles.

Pāhoehoe flows enter the Pacific Ocean on Hawaii island.

The Thurston lava tube on Hawaii island, now a tourist attraction in the Hawaii Volcanoes National Park.

Distribution

Shield volcanoes are found worldwide. They can form over hotspots (points where magma from below the surface wells up), such as the Hawaiian–Emperor seamount chain and the Galápagos Islands, or over more conventional rift zones, such as the Icelandic shields and the shield volcanoes of East Africa. Many shield volcanoes are found in ocean basins, such as Tamu Massif, the world's largest, although they can be found inland as well—East Africa being one example of this.

Hawaiian Islands

The largest and most prominent shield volcano chain in the world is the Hawaiian Islands, a chain of hotspot volcanoes in the Pacific Ocean. The Hawaiian volcanoes are characterized by frequent rift eruptions, their large size (thousands of km^3 in volume), and their rough, decentralized shape. Rift zones are a prominent feature on these volcanoes, and account for their seemingly random volcanic structure. They are fueled by the movement of the Pacific Plate over the Hawaii hotspot, and form a long chain of volcanoes, atolls, and seamounts 2,600 km (1,616 mi) long with a total volume of over 750,000 km^3 (179,935 cu mi). The chain contains at least 43 major volcanoes, and Meiji Seamount at its terminus near the Kuril–Kamchatka Trench is 85 million years old. The volcanoes follow a distinct evolutionary pattern of growth and death.

The chain includes the second largest volcano on Earth, Mauna Loa, which stands 4,170 m (13,680 ft) above sea level and reaches a further 13 km (8 mi) below the waterline and into the crust, approximately 80,000 km^3 (19,000 cu mi) of rock. Kīlauea, meanwhile, is one of the most active volcanoes on Earth, with the current ongoing eruption having begun in January 1983.

Galápagos Islands

The Galápagos Islands are an isolated set of volcanoes, consisting of shield volcanoes and lava plateaus, located 1,200 km (746 mi) west of Ecuador. They are driven by the Galápagos hotspot, and are between approximately 4.2 million and 700,000 years of age. The largest island, Isabela Island, consists of six coalesced shield volcanoes, each delineated by a large summit caldera. Española, the oldest island, and Fernandina, the youngest, are also shield volcanoes, as are most of the other islands in the chain. The Galápagos Islands are perched on a large lava plateau known as the Galápagos Platform. This platform creates a shallow water depth of 360 to 900 m (1,181 to 2,953 ft) at the base of the islands, which stretch over a 174 mi (280 km)-long diameter. Since Charles Darwin's visit to the islands in 1835 during the Second voyage of HMS Beagle, there have been over 60 recorded eruptions in the islands, from six different shield volcanoes. Of the 21 emergent volcanoes, 13 are considered active.

Blue Hill is a shield volcano on the south western part of Isabela Island in the Galápagos Islands and is one of the most active in the Galapagos, with the last eruption between May and June 2008. The Geophysics Institute at the National Polytechnic School in Quito houses an international team of seismologists and volcanologists whose responsibility is to monitor Ecuadors numerous active volcanoes in the Andean Volcanic Belt and the Galapagos Islands. La Cumbre is an active shield volcano on Fernandina Island in the Galapagos that has been erupting since April 11, 2009.

The Galápagos islands are geologically young for such a big chain, and the pattern of their rift zones follows one of two trends, one north-northwest, and one east–west. The composition of the lavas of the Galápagos shields are strikingly similar to those of the Hawaiian volcanoes. Curiously, they do not form the same volcanic "line" associated with most hotspots. They are not alone in this regard; the Cobb–Eickelberg Seamount chain in the North Pacific is another example of such a delineated chain. In addition, there is no clear pattern of age between the volcanoes, suggesting a complicated, irregular pattern of creation. How exactly the islands were formed remains a geological mystery, although several theories have been fronted.

Iceland

Skjaldbreiður, Iceland, is eponymous for shield volcanoes.

Another major center of shield volcanic activity is Iceland. Located over the Mid-Atlantic Ridge, a divergent tectonic plate in the middle of the Atlantic Ocean, Iceland is the site of about 130 volcanoes of various types. Icelandic shield volcanoes are generally of Holocene age, between 5,000 and 10,000 years old, except for the island of Surtsey, a Surtseyan shield. The volcanoes are also very narrow in distribution, occurring in two bands in the West and North Volcanic Zones. Like Hawaiian volcanoes, their formation initially begins with several eruptive centers before centralizing and concentrating at a single point. The main shield then forms, burying the smaller ones formed by the early eruptions with its lava.

Icelandic shields are mostly small (~15 km^3 (4 cu mi)), symmetrical (although this can affected by surface topography), and characterized by eruptions from summit calderas. They are composed of either tholeiitic olivine or picritic basalt. The tholeiitic shields tend to be wider and shallower than the picritic shields. They do not follow the pattern of caldera growth and destruction that other shield volcanoes do; caldera may form, but they generally do not disappear.

East Africa

East Africa is the site of volcanic activity generated by the development of the East African Rift, a developing plate boundary in Africa, and from nearby hotspots. Some volcanoes interact with both. Shield volcanoes are found near the rift and off the coast of Africa, although stratovolcanoes are more common. Although sparsely studied, the fact that all of its volcanoes are of Holocene age reflects how young the volcanic center is. One interesting characteristic of East African volcanism is a penchant for the formation of lava lakes; these semi-permanent lava bodies, extremely rare elsewhere, form in about nine percent of African eruptions.

The most active shield volcano in Africa is Nyamuragira. Eruptions at the shield volcano are generally centered within the large summit caldera or on the numerous fissures and cinder cones on the volcano's flanks. Lava flows from the most recent century extend down the flanks more than 30 km (19 mi) from the summit, reaching as far as Lake Kivu. Erta Ale in Ethiopia is another active shield volcano, and one of the few places in the world with a permanent lava lake, which has been active since at least 1967, and possibly since 1906. Other volcanic centers include Menengai, a massive shield caldera, and Mount Marsabit, near the town of Marsabit.

Extraterrestrial Volcanoes

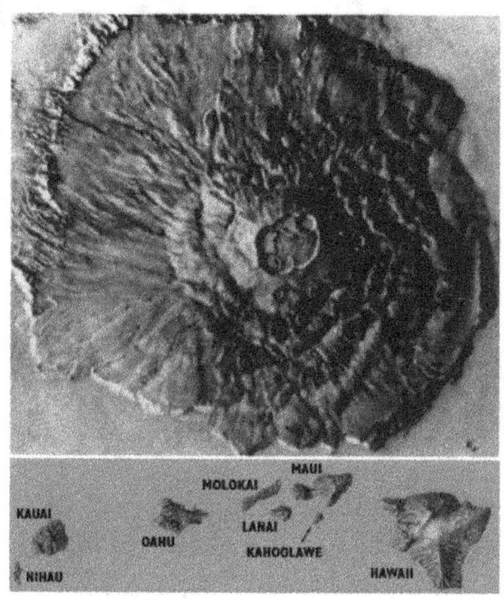

Scaled image showing Olympus Mons, top, and the Hawaiian island chain, bottom. Martian volcanoes are far larger than those found on Earth.

Volcanoes are not limited to Earth; they can exist on any rocky planet or moon large or active enough to have a molten core, and since probes were first launched in the 1960s, volcanoes have been found across the solar system. Shield volcanoes and volcanic vents have been found on Mars, Venus, and Io; cryovolcanoes on Triton; and subsurface hotspots on Europa.

The volcanoes of Mars are very similar to the shield volcanoes on Earth. Both have gently sloping flanks, collapse craters along their central structure, and are built of highly fluid lavas. Volcanic features on Mars were observed long before they were first studied in detail during the 1976–1979 Viking mission. The principal difference between the volcanoes of Mars and those on Earth is in terms of size; Martian volcanoes range in size up to 14 mi (23 km) high and 370 mi (595 km) in diameter, far larger than the 6 mi (10 km) high, 74 mi (119 km) wide Hawaiian shields. The highest of these, Olympus Mons, is the tallest known mountain on any planet in the solar system.

Venus also has over 150 shield volcanoes which are much flatter, with a larger surface area than those found on Earth, some having a diameter of more than 700 km (430 mi). Although the majority of these are long extinct it has been suggested, from observations by the Venus Express spacecraft, that many may still be active.

Cryovolcanoes

A cryovolcano (sometimes informally called an ice volcano) is a type of volcano that erupts volatiles such as water, ammonia or methane, instead of molten rock. Collectively referred to as cryomagma, cryolava or ice-volcanic melt, these substances are usually liquids and can form plumes, but can also be in vapour form. After eruption, cryomagma is expected to condense to a solid form when exposed to the very low surrounding temperature. Cryovolcanoes may potentially form on icy moons and other objects with abundant water past the Solar System's snow line (such as Pluto). A number of features have been identified as possible cryovolcanoes on Pluto, Titan and Ceres. In addition, although they are not known to form volcanoes, ice geysers have been observed on Enceladus and potentially Triton.

One potential energy source on some solar system bodies for melting ices and producing cryovolcanoes is tidal friction. Translucent deposits of frozen materials create a subsurface greenhouse effect that would accumulate the required heat.

Signs of past warming of the Kuiper belt object Quaoar have led scientists to speculate that it exhibited cryovolcanism in the past. Radioactive decay could provide the energy necessary for such activity, as cryovolcanoes can emit water mixed with ammonia, which would melt at 180 K (−95 °C) and create an extremely cold liquid that would flow out of the volcano.

Observations

Plumes of Enceladus, feeding Saturn's E Ring, seem to arise from the "Tiger Stripes" near the south pole.

On November 27, 2005, Cassini photographed geysers on the south pole of Enceladus. Indirect evidence of cryovolcanic activity was later observed on several other icy moons of the Solar System, including Europa, Titan, Ganymede, and Miranda. Cassini has observed several features thought to be cryovolcanoes on Titan, notably Doom Mons with adjacent Sotra Patera, a feature regarded as "the very best evidence, by far, for volcanic topography anywhere documented on an icy satellite." Cryovolcanism is one process hypothesized to be a significant source of the methane found in Titan's atmosphere.

In 2007, observations by the Gemini Observatory showing patches of ammonia hydrates and water crystals on the surface of Pluto's moon Charon suggested the presence of active cryovolcanoes or cryogeysers. Subsequent observations by New Horizons in 2015 found that Charon has a youthful surface, supporting this idea. Pluto itself has two features that have been identified as possible cryovolcanoes, being mountains with indented peaks.

In 2015, two distinct bright spots inside a crater of the dwarf planet Ceres were imaged by the Dawn spacecraft, leading to speculation about a possible cryovolcanic origin.

In September 2016, NASA JPL and NASA Goddard scientists released findings that large Ahuna Dome on Ceres is a "volcanic dome unlike any seen elsewhere in the solar system. The large mountain is likely volcanic in nature. Specifically, it would be a cryovolcano -- a volcano that erupts a liquid made of volatiles such as water, instead of silicates. The only known example of a cryovolcano that potentially formed from a salty mud mix, and that formed in the geologically recent past." In addition, at least some of Ceres' well known bright spots (notably including the ones in Occator crater) are likely also cryovolcanic in origin. A study published in March 2017 suggests that Occator's most recent large eruption occurred about 4 million years ago and thus that Ceres may still be active.

Cryovolcanism on Enceladus

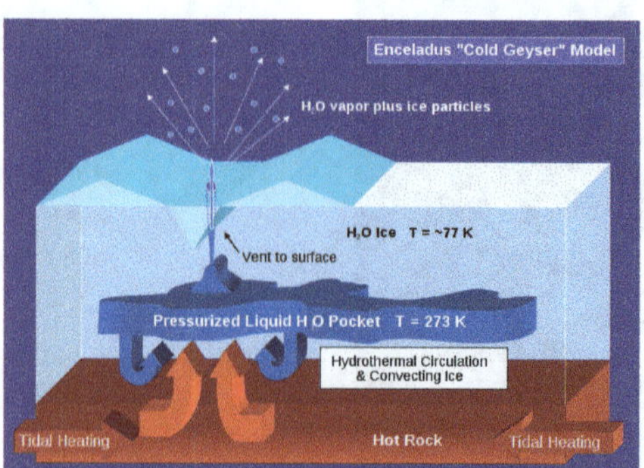

One possible scheme for Enceladus's cryovolcanism.

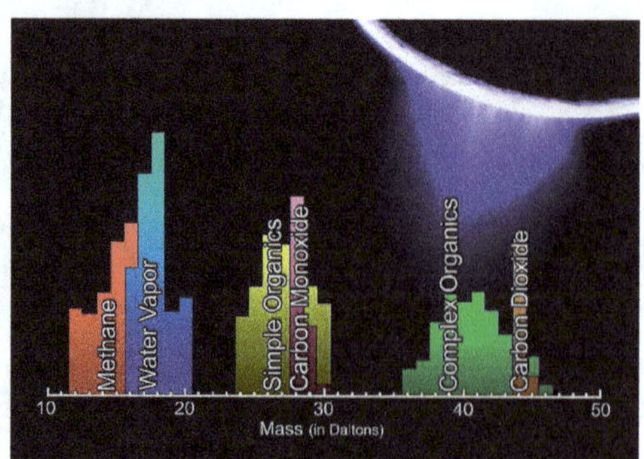

Chemical composition of the plumes of Enceladus.

Subglacial Volcanoes

A subglacial volcano, also known as a glaciovolcano, is a volcanic form produced by subglacial eruptions or eruptions beneath the surface of a glacier or ice sheet which is then melted into a lake by the rising lava. Today they are most common in Iceland and Antarctica; older formations of this type are found also in British Columbia and Yukon Territory, Canada.

During the eruption, the heat of the lava from the subglacial volcano melts the overlying ice. The water quickly cools the lava, resulting in pillow lava shapes similar to those of underwater volcanoes. When the pillow lavas break off and roll down the volcano slopes, pillow breccia, tuff breccia, and hyaloclastite form. The meltwater may be released from below the ice as happened in Iceland in 1996 when the Grímsvötn caldera erupted, melting 3 km^3 of ice and giving rise to a large glacial lake outburst flood.

The shape of subglacial volcanoes tends to be quite characteristic and unusual, with a flattened top and steep sides supported against collapse by the pressure of the surrounding ice and meltwater. If the volcano eventually melts completely through the ice layer, then horizontal lava flows are deposited, and the top of the volcano assumes a nearly level form. However, if significant amounts of lava are later erupted subaerially, then the volcano may assume a more conventional shape. In Canada the volcanos have been known to form both conical and nearly level shapes. The more distinctly flat-topped, steep-sided subglacial volcanoes are called tuyas, named after Tuya Butte in northern British Columbia by Canadian geologist Bill Mathews in 1947. In Iceland, such volcanoes are also known as table mountains.

Jökulhlaups

Subglacial eruptions often cause jökulhlaups or great floods of water. In November 1996 the Grímsvötn Volcano beneath the Vatnajökull ice sheet erupted and caused a Jökulhlaup that affected more than 750 km^2 (290 sq mi) and destroyed or severely damaged several bridges. During the ice ages, such floods from Lake Missoula were estimated to have discharges exceeding 17×10^6 m^3/s (4.5×10^9 gal/s) and covered a third of eastern Washington state. Sonia Esperanca, program director in the National Science Foundation commented on the danger of subglacial volcanoes: "When an ice-covered volcano erupts, the interplay among molten magma, ice and meltwater can have catastrophic results."

Antarctica Eruption

In January, 2008, the British Antarctic Survey (Bas) scientists led by Hugh Corr and David Vaughan, reported that 2,200 years ago, a volcano erupted under the Antarctica ice sheet (based on airborne survey with radar images). The biggest eruption in the last

10,000 years, the volcanic ash was found deposited on the ice surface under the Hudson Mountains, close to Pine Island Glacier.

On Mars

Many scientists believe that liquid water exists many kilometers below the surface of Mars, but at this point in time it is impossible to drill to those depths with the rovers in existence. Meredith Payne and Jack Farmer of Arizona State University have studied images from the Viking and Mars Orbiter cameras in search of possible sub-glacial volcanoes that could carry microbes to the surface.

Ice Cores

It is possible to track catastrophic subglacial volcano eruptions in time with the analysis of ice cores such as the Vostok core. Subglacial volcanic eruptions are identified by layers of high concentrations of NO^{-}_{3} and SO^{2-}_{4}.

Ice Volcanoes

An ice volcano is a conical mound of ice formed over a terrestrial lake via the eruption of water and slush through an ice shelf. The process is wave-driven, with wind providing the energy for the waves to cut through the ice and form the volcanoes. After being ejected into the atmosphere, the liquid water and slush freeze and fall back to the surface, growing the formation. Ice may also be erupted. The phenomenon is most often observed along the southern coast of Lake Erie and Lake Ontario, when the temperature is below freezing and the wind blows onshore with a velocity of at least 25 mph (40 km/h). They are known to reduce coastal erosion there. The formations are temporary — they are frequently destroyed by storms and warm weather, and once the lake wholly freezes over, eruptions are no longer possible.

There is no consensus name for this phenomenon. Due to its similarity to volcanism and particularly cryovolcanism, the term “ice volcano” is frequently used, but it remains controversial. Unlike geysers and related structures, ice volcanoes are not hydrothermal.

The uplifts may attract a number of visitors, but they are dangerous, and experts warn that people may fall through the ice or slip into the cold lake. Ice volcanoes are used by snowy owls as hunting platforms to search for waterfowl.

Formation

These features are distinct from pressure ridges, which are uplifts formed by the compression of ice against a shoreline or another floe. Instead, ice volcanoes are created by waves colliding with irregularities at the edge of an ice sheet. The abnormalities concentrate the wave energy in a small area, where the ice is eroded to form a V-shaped channel. Spray, ice, and slush splashing out of the feature create a volcanic cone at the channel's shoreward end. This process takes only a few hours. The lakeward end of the channel may then by sealed by ice, but the volcano may continue to erupt. A wave amplitude of at least one metre (3 ft 3 in) is needed to induce eruptions, so ice volcanoes are rarely active without storm-force winds. Formation near land is suppressed by reefs and shoals, which absorb the wave energy needed for the phenomenon. Nonetheless, they may produce larger cones further out at sea, where the greater depth makes this possible. Formation is more thoroughly suppressed by powerful storms, which erode the ice too fast for mound creation.

One type of ice volcano, known as a cold spot, does not require waves to break against the edge of an ice shelf. Instead, water and slush erupt through a region of weak ice near the coast and form a mound. This is analogous to a geological hotspot.

Appearance and Eruptions

Landfast ice is required, so the volcanoes normally form near land. They are found in successive rows, and within one row, the features usually have equal height and spacing. However, when comparing two rows, the height and spacing may be drastically different. Ice volcanoes range in height from less than one meter to ten meters, with the largest ones located far from the shore. Eruptions over ten meters high have been observed, but it is believed that the height of the eruptions are proportional to the size of the mounds. A single eruption may increase the height of the volcano by several centimeters. When an eruption occurs above 0 °C, however, the water erodes the uplift instead of expanding it. Spacing is determined by the amplitude and direction of the waves. In general, the appearance and number of ice volcanoes change considerably between winters.

Different types of ice volcanoes have been compared to shield volcanoes and stratovolcanoes. They are noted for their symmetry. Cold spot volcanoes are particularly symmetrical, but their eruption has not been observed.

Mud Volcanoes

A mud volcano or mud dome is a landform created by the eruption of mud or slurries, water and gases. Several geological processes may cause the formation of mud volcanoes. Mud volcanoes are not true igneous volcanoes as they do not produce lava and are not necessarily driven by magmatic activity. The Earth continuously exudes a mud-like substance, which may sometimes be referred to as a "mud volcano". Mud volcanoes may range in size from merely 1 or 2 meters high and 1 or 2 meters wide, to 700 meters high and 10 kilometers wide. Smaller mud exudations are sometimes referred to as mud-pots.

The mud produced by mud volcanoes is mostly formed as hot water, which has been heated deep below the Earth's surface, begins to mix and blend with subterranean mineral deposits, thus creating the mud slurry exudate. This material is then forced upwards through a geological fault or fissure due to local subterranean pressure imbalances. Mud volcanoes are associated with subduction zones and about 1100 have been identified on or near land. The temperature of any given active mud volcano generally remains fairly steady and is much lower than the typical temperatures found in igneous volcanoes. Mud volcano temperatures can range from near 100 °C (212 °F) to occasionally 2 °C (36 °F), some being used as popular "mud baths."

About 86% of the gas released from these structures is methane, with much less carbon dioxide and nitrogen emitted. Ejected materials are most often a slurry of fine solids suspended in water that may contain a mixture of salts, acids and various hydrocarbons. Possible mud volcanoes have been identified on Mars.

A mud volcano may be the result of a piercement structure created by a pressurized mud diapir that breaches the Earth's surface or ocean bottom. Their temperatures may be as low as the freezing point of the ejected materials, particularly when venting is associated with the creation of hydrocarbon clathrate hydrate deposits. Mud volcanoes are often associated with petroleum deposits and tectonic subduction zones and orogenic belts; hydrocarbon gases are often erupted. They are also often associated with lava volcanoes; in the case of such close proximity, mud volcanoes emit incombustible gases including helium, whereas lone mud volcanoes are more likely to emit methane.

Approximately 1,100 mud volcanoes have been identified on land and in shallow water. It has been estimated that well over 10,000 may exist on continental slopes and abyssal plains.

Features

- Gryphon: steep-sided cone shorter than 3 meters that extrudes mud.
- Mud cone: high cone shorter than 10 meters that extrudes mud and rock fragments.
- Scoria cone: cone formed by heating of mud deposits during fires.
- Salse: water-dominated pools with gas seeps.
- Spring: water-dominated outlets smaller than 0.5 metres.
- Mud Shield.

Emissions

Most liquid and solid material is released during eruptions, but seeps occur during dormant periods. The mud is rich in halite (rock salt). First-order estimates of mud volcano emissions have been made (1 Tg = 1 million metric tonnes).

- 2002: L. I. Dimitrov estimated that 10.2–12.6 Tg/yr of methane is released from onshore and shallow offshore mud volcanoes.
- 2002: Etiope and Klusman estimated at least 1–2 and as much as 10–20 Tg/yr of methane may be emitted from onshore mud volcanoes.
- 2003: Etiope, in an estimate based on 120 mud volcanoes: "The emission results to be conservatively between 5 and 9 Tg/yr, that is 3–6% of the natural methane sources officially considered in the atmospheric methane budget. The total geologic source, including MVs (this work), seepage from seafloor (Kvenvolden et al., 2001), microseepage in hydrocarbon-prone areas and geothermal sources (Etiope and Klusman, 2002), would amount to 35–45 Tg/yr."
- 2003: analysis by Milkov et al. suggests that the global gas flux may be as high as 33 Tg/yr (15.9 Tg/yr during quiescent periods plus 17.1 Tg/yr during eruptions). Six teragrams per year of greenhouse gases are from onshore and shallow offshore mud volcanoes. Deep-water sources may emit 27 Tg/yr. Total may be 9% of fossil CH_4 missing in the modern atmospheric CH_4 budget, and 12% in the preindustrial budget.
- 2003: Alexei Milkov estimated approximately 30.5 Tg/yr of gases (mainly methane and CO_2) may escape from mud volcanoes to the atmosphere and the ocean.
- 2003: Achim J. Kopf estimated 1.97×10^{11} to 1.23×10^{14} m³ of methane is released by all mud volcanoes per year, of which 4.66×10^{7} to 3.28×10^{11} m³ is from surface volcanoes. That converts to 141–88,000 Tg/yr from all mud volcanoes, of which 0.033–235 Tg is from surface volcanoes.

Caldera

A caldera is a large cauldron-like hollow that forms shortly after the emptying of a magma chamber/reservoir in a volcanic eruption. When large volumes of magma are erupted over a short time, structural support for the rock above the magma chamber is lost. The ground surface then collapses downward into the emptied or partially emptied magma chamber, leaving a massive depression at the surface (from one to dozens of kilometers in diameter). Although sometimes described as a crater, the feature is actually a type of sinkhole, as it is formed through subsidence and collapse rather than an explosion or impact. Only seven caldera-forming collapses are known to have occurred since 1900, most recently at Bárðarbunga volcano, Iceland in 2014.

Caldera Formation

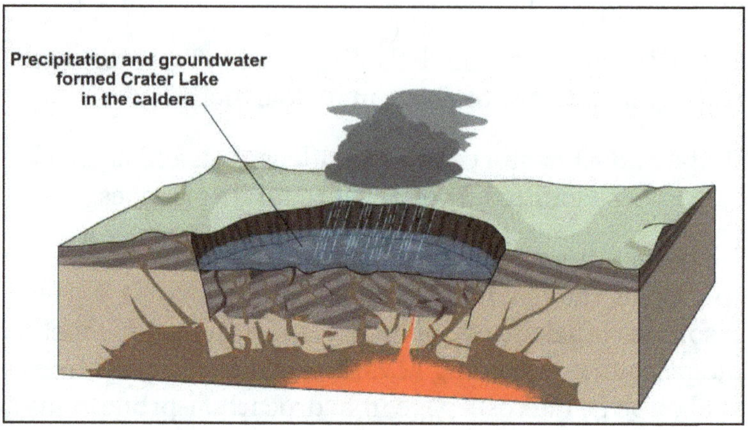

Formation of a Crater Lake-type caldera.

A collapse is triggered by the emptying of the magma chamber beneath the volcano, sometimes as the result of a large explosive volcanic eruption, but also during effusive eruptions on the flanks of a volcano or in a connected fissure system. If enough magma is ejected, the emptied chamber is unable to support the weight of the volcanic edifice above it. A roughly circular fracture, the "ring fault", develops around the edge of the chamber. Ring fractures serve as feeders for fault intrusions which are also known as ring dikes. Secondary volcanic vents may form above the ring fracture. As the magma chamber empties, the center of the volcano within the ring fracture begins to collapse. The collapse may occur as the result of a single cataclysmic eruption, or it may occur in stages as the result of a series of eruptions. The total area that collapses may be hundreds or thousands of square kilometers.

Landsat image of Lake Toba, on the island of Sumatra, Indonesia (100 km/62 mi long and 30 km/19 mi wide, one of the world's largest calderas). A resurgent dome formed the island of Samosir.

Mineralization in Calderas

Some calderas are known to host rich ore deposits. Metal-rich fluids can circulate through the caldera, forming hydrothermal ore deposits of metals such as lead, silver, gold, mercury, lithium and uranium. One of the world's best-preserved mineralized calderas is the Sturgeon Lake Caldera in northwestern Ontario, Canada, which formed during the Neoarchean era about 2,700 million years ago.

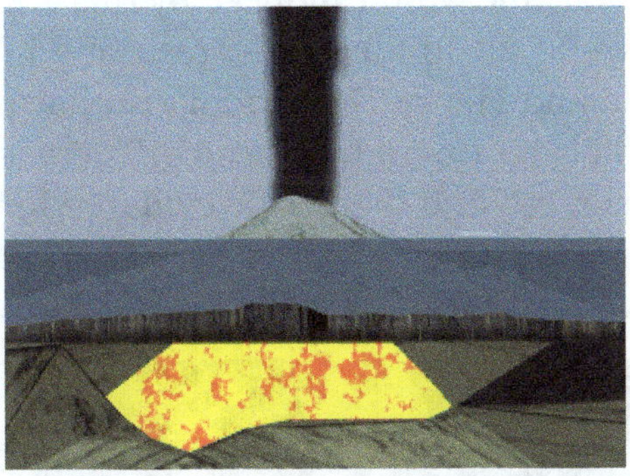

Caldera formation under water.

Types of Caldera

Explosive Caldera Eruptions

If the magma is rich in silica, the caldera is often filled in with ignimbrite, tuff, rhyolite, and other igneous rocks. Silica-rich magma has a high viscosity, and therefore does not flow easily like basalt. As a result, gases tend to become trapped at high pressure within the magma. When the magma approaches the surface of the Earth, the rapid off-loading of overlying material causes the trapped gases to decompress rapidly, thus triggering explosive destruction of the magma and spreading volcanic ash over wide areas. Further lava flows may be erupted.

If volcanic activity continues, the center of the caldera may be uplifted in the form of a resurgent dome such as is seen at Cerro Galán, Lake Toba, Yellowstone, etc., by subsequent intrusion of magma. A silicic or rhyolitic caldera may erupt hundreds or even thousands of cubic kilometers of material in a single event. Even small caldera-forming eruptions, such as Krakatoa in 1883 or Mount Pinatubo in 1991, may result in significant local destruction and a noticeable drop in temperature around the world. Large calderas may have even greater effects.

When Yellowstone Caldera last erupted some 650,000 years ago, it released about 1,000 km^3 of material (as measured in dense rock equivalent (DRE)), covering a substantial part of North America in up to two metres of debris. By comparison, when Mount St. Helens erupted in 1980, it released ~1.2 km^3 (DRE) of ejecta. The ecological effects of the eruption of a large caldera can be seen in the record of the Lake Toba eruption in Indonesia.

Toba

About 74,000 years ago, this Indonesian volcano released about 2,800 cubic kilometres

(670 cu mi) dense-rock equivalent of ejecta. This was the largest known eruption during the ongoing Quaternary period (the last 2.6 million years) and the largest known explosive eruption during the last 25 million years. In the late 1990s, anthropologist Stanley Ambrose proposed that a volcanic winter induced by this eruption reduced the human population to about 2,000–20,000 individuals, resulting in a population bottleneck. More recently, Lynn Jorde and Henry Harpending proposed that the human species was reduced to approximately 5,000-10,000 people. There is no direct evidence, however, that either theory is correct, and there is no evidence for any other animal decline or extinction, even in environmentally sensitive species. There is evidence that human habitation continued in India after the eruption.

Satellite photograph of the summit caldera on Fernandina Island in the Galápagos archipelago.

Eruptions forming even larger calderas are known, especially La Garita Caldera in the San Juan Mountains of Colorado, where the 5,000 cubic kilometres (1,200 cu mi) Fish Canyon Tuff was blasted out in eruptions about 27.8 million years ago.

Oblique aerial photo of Nemrut Caldera, Van Lake, Eastern Turkey.

At some points in geological time, rhyolitic calderas have appeared in distinct clusters. The remnants of such clusters may be found in places such as the San Juan Mountains of Colorado (formed during the Oligocene, Miocene, and Pliocene epochs) or the Saint Francois Mountain Range of Missouri (erupted during the Proterozoic eon).

Non-explosive Calderas

Some volcanoes, such as the large shield volcanoes Kīlauea and Mauna Loa on the island of Hawaii, form calderas in a different fashion. The magma feeding these volcanoes is basalt, which is silica poor. As a result, the magma is much less viscous than the magma of a rhyolitic volcano, and the magma chamber is drained by large lava flows rather than by explosive events. The resulting calderas are also known as subsidence calderas and can form more gradually than explosive calderas. For instance, the caldera atop Fernandina Island collapsed in 1968 when parts of the caldera floor dropped 350 metres (1,150 ft).

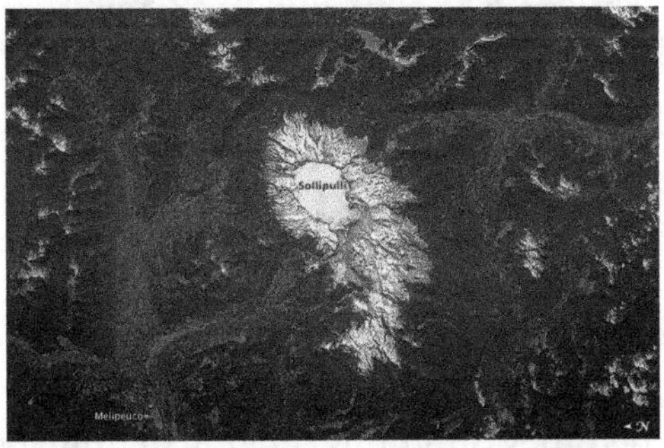

Sollipulli Caldera, located in central Chile near the border with Argentina, filled with ice. The volcano is in the southern Andes Mountains within Chile's Parque Nacional Villarica.

Extraterrestrial Calderas

Since the early 1960s, it has been known that volcanism has occurred on other planets and moons in the Solar System. Through the use of manned and unmanned spacecraft, volcanism has been discovered on Venus, Mars, the Moon, and Io, a satellite of Jupiter. None of these worlds have plate tectonics, which contributes approximately 60% of the Earth's volcanic activity (the other 40% is attributed to hotspot volcanism). Caldera structure is similar on all of these planetary bodies, though the size varies considerably. The average caldera diameter on Venus is 68 km (42 mi). The average caldera diameter on Io is close to 40 km (25 mi), and the mode is 6 km (3.7 mi); Tvashtar Paterae is likely the largest caldera with a diameter of 290 km (180 mi). The average caldera diameter on Mars is 48 km (30 mi), smaller than Venus. Calderas on Earth are the smallest of all planetary bodies and vary from 1.6–80 km (1–50 mi) as a maximum.

The Moon

The Moon has an outer shell of low-density crystalline rock that is a few hundred kilometers thick, which formed due to a rapid creation. The craters of the Moon have been well preserved through time and were once thought to have been the result of extreme volcanic activity, but actually were formed by meteorites, nearly all of which took place in the first few hundred million years after the Moon formed. Around 500 million years afterward, the Moon's mantle was able to be extensively melted due to the decay of radioactive elements. Massive basaltic eruptions took place generally at the base of large impact craters. Also, eruptions may have taken place due to a magma reservoir at the base of the crust. This forms a dome, possibly the same morphology of a shield volcano where calderas universally are known to form. Although caldera-like structures are rare on the Moon, they are not completely absent. The Compton-Belkovich Volcanic Complex on the far side of the Moon is thought to be a caldera, possibly an ash-flow caldera.

Mars

The volcanic activity of Mars is concentrated in two major provinces: Tharsis and Elysium. Each province contains a series of giant shield volcanoes that are similar to what we see on Earth and likely are the result of mantle hot spots. The surfaces are dominated by lava flows, and all have one or more collapse calderas. Mars has the largest volcano in the Solar System, Olympus Mons, which is more than three times the height of Mount Everest, with a diameter of 520 km (323 miles). The summit of the mountain has six nested calderas.

Venus

Because there is no plate tectonics on Venus, heat is mainly lost by conduction through the lithosphere. This causes enormous lava flows, accounting for 80% of Venus' surface area. Many of the mountains are large shield volcanoes that range in size from 150–400 km (95–250 mi) in diameter and 2–4 km (1.2–2.5 mi) high. More than 80 of these large shield volcanoes have summit calderas averaging 60 km (37 mi) across.

Io

Io, unusually, is heated by solid flexing due to the tidal influence of Jupiter and Io's orbital resonance with neighboring large moons Europa and Ganymede, which keeps its orbit slightly eccentric. Unlike any of the planets mentioned, Io is continuously volcanically active. For example, the NASA Voyager 1 and Voyager 2 spacecraft detected nine erupting volcanoes while passing Io in 1979. Io has many calderas with diameters tens of kilometers across.

Supervolcanoes

A supervolcano is any volcano capable of producing a volcanic eruption with an ejecta volume greater than 1,000 km3 (240 cu mi). This is thousands of times larger than normal volcanic eruptions. Supervolcanoes are on a much bigger scale than other volcanoes. Unlike composite volcanoes, with their steep sides, they are difficult to spot. They are typically depresssions in the ground. The calderas are so large they can be seen from space. They have been identified in Indonesia, in New Zealand, in South America and an extinct one in Glen Coe in the UK.

Supervolcanoes can occur when magma in the mantle rises into the crust from a hotspot but is unable to break through the crust, and pressure builds in a large and growing magma pool until the crust is unable to contain the pressure (this is the case for the Yellowstone Caldera). They can also form at convergent plate boundaries (for example, Toba). Although there are only a handful of Quaternary supervolcanoes, supervolcanic eruptions typically cover huge areas with lava and volcanic ash and cause a long-lasting change to weather (such as the triggering of a small ice age) sufficient to threaten species with extinction.

Characteristics of a Supervolcano

Supervolcanoes are not mountains – they form depressions within the Earth's crust. They begin with a column of magma rising through a vent into the Earth's crust. The magma gets stuck and pools, melting the rock around for thousands of years. Over thousands of years the pressure builds up and when the eruption eventually happens it drains the magma lake and the land above collapse down over, creating a caldera.

The stages in the creation of a supervolcano caldera

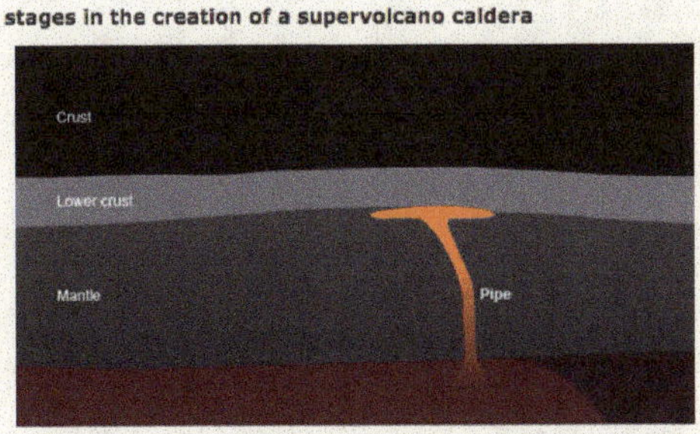

Magma cannot escape to the surface and collects under the lower crust.

The stages in the creation of a supervolcano caldera

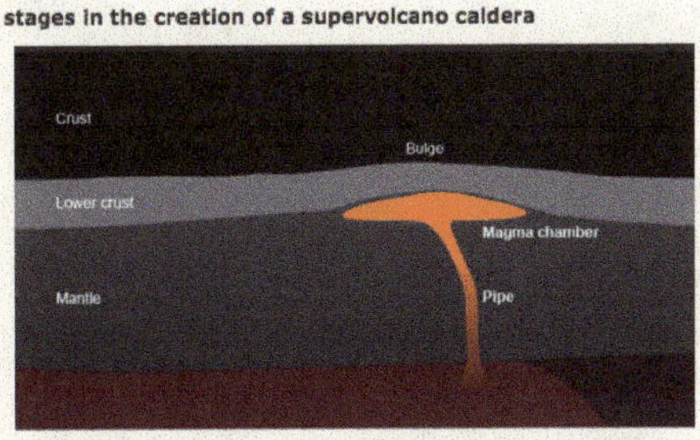

An 'uplifted bulge' begins to form under the lower crust as the magma chamber enlarges.

The stages in the creation of a supervolcano caldera

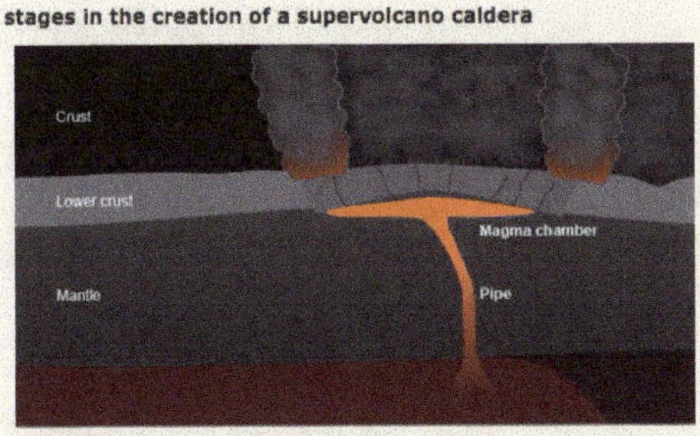

Cracks appear on the surface. Gas and ash erupt from the magma chamber through these cracks.

The stages in the creation of a supervolcano caldera

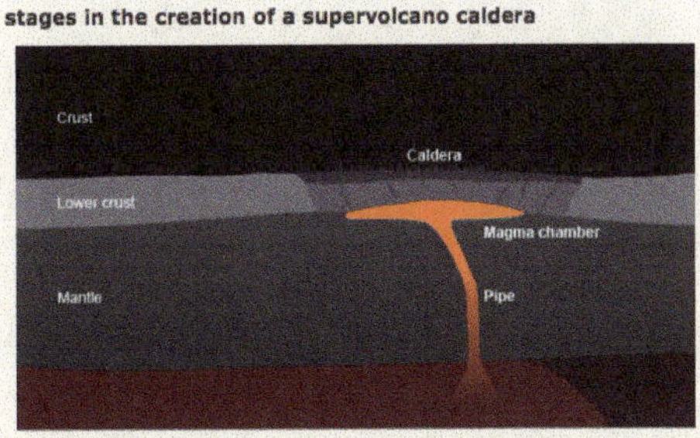

The magma chamber collapses and a depression is formed. This is called a caldera.

Effects of Supervolcanoes

Supervolcanoes are eruptions and explosions of catastrophic proportions, on the Volcano Explosivity Index (VEI) supervolcanoes are an 8 on a scale that runs from 1 to 8. Each leap up the scale represents an increase of explosive scale of 10 times the power. Mount St Helens was VEI 5. There have been estimates that a major eruption of Yellowstone would result in at least 1 m of ash being deposited across the UK.

References

- "Garibaldi volcanic belt: Garibaldi Lake volcanic field". Catalogue of Canadian volcanoes. Geological Survey of Canada. 2009-04-01. Archived from the original on June 26, 2009. Retrieved 2010-06-27
- Submarine-volcano: basicplanet.com, Retrieved 3 April, 2019
- Stewart, Emily M.; Head, James W. (1 August 2001). "Ancient Martian volcanoes in the Aeolis region: New evidence from MOLA data". Journal of Geophysical Research. 106 (E8): 17505. doi:10.1029/2000JE001322
- Whatisasupervolcano, topic: geography.learnontheinternet.co.uk, Retrieved 4 May, 2019
- Media Relations Office: Cassini Imaging Central Laboratory For Operations (2009). "Cassini Finds Hydrocarbon Rains May Fill The Lakes". Space Science Institute, Boulder, Colorado. Retrieved 4 March 2015

3
Volcanology

The study of volcanoes, lava, magma and other geological phenomena related to volcanoes is known as volcanology. This field of geology analyzes all the major aspects of volcanoes such as volcanic arcs, volcanic belt, volcanic gas, lava balloons, etc. All these areas of study within volcanology have been carefully analyzed in this chapter.

Volcanology, also spelled vulcanology is a discipline of the geologic sciences that is concerned with all aspects of volcanic phenomena.

Volcanology deals with the formation, distribution, and classification of volcanoes as well as with their structure and the kinds of materials ejected during an eruption (such as pyroclastic flows, lava, dust, ash, and volcanic gases). It also involves research on the relationships between volcanic eruptions and other large-scale geologic processes such as plate tectonics, mountain building, and earthquakes. One of the chief objectives of this research is determining the nature and causes of volcanic eruptions for the purpose of forecasting their occurrence. Another practical concern of volcanology is securing data that may aid in locating commercially valuable deposits of ores, particularly those of certain sulfide minerals.

Interest in volcanic phenomena extends back to ancient times. The eruption of Mount Vesuvius in AD 79 was recorded in considerable detail by Pliny the Younger. Studies of volcanoes, however, were not conducted systematically until the 19th century. Since that time volcanology has become an important branch of physical geology. Specialists in the field, using the principles and methods of geophysics and geochemistry and the tools of seismology and geodesy, have obtained much knowledge of processes that occur within the Earth's interior.

Volcanic Arcs

All subduction zones have, at some distance in from the edge of the upper plate, arcs or chains of composite cone volcanoes. The subducting plate, as it goes down deep into the mantle, releases water. This changes the chemistry of the already hot rocks in the mantle and causes them to melt, forming magma. The magma is less dense than the solid rocks around it, so it rises upward, culminating in volcanic eruptions at the earth's surface.

The volcanic arc at an ocean-continent subduction zone is not only a chain of volcanoes. The stress of plate convergence compresses the crust there, causing it to thicken through a combination of folds and thrust faults. Igneous intrusions and volcanic eruptions also thicken the crust there. Deep within the crust, the igneous intrusions solidify into batholiths of rocks such as granite, and the pre-existing rocks that are intruded by the batholiths are regionally metamorphosed into new rocks. The result is a high mountain range with granitic and metamorphic rock at its core, folded and faulted sedimentary and volcanic around its margins, and a chain of composite cone volcanoes distributed along the crest of the range.

Volcanic Belt

A volcanic belt is a large volcanically active region. Other terms are used for smaller areas of activity, such as volcanic fields. Volcanic belts are found above zones of unusually high temperature (700-1400 °C) where magma is created by partial melting of solid material in the Earth's crust and upper mantle. These areas usually form along tectonic plate boundaries at depths of 10–50 km. For example, volcanoes in Mexico and western North America are mostly in volcanic belts, such as the Trans-Mexican Volcanic Belt that extends 900 km from west to east across central-southern Mexico and the Northern Cordilleran Volcanic Province in western Canada.

The deeply deformed and eroded remnants of ancient volcanic belts are found in volcanically inactive regions such as the Canadian Shield. It contains over 150 volcanic belts (now deformed and eroded down to nearly flat plains) that range from 600 to 1200 million years old. These are zones of variably metamorphosed mafic to ultramafic volcanic sequences with associated sedimentary rocks that form what are known as greenstone belts. They are thought to have formed at ancient oceanic spreading centers and island arc terranes. The Abitibi greenstone belt in Ontario and Quebec, Canada is one of the world's largest greenstone belts.

Volcanic belts are similar to a mountain range, but the mountains within the mountain range are volcanoes, not mountains that are formed by faulting and folding by the collision of tectonic plates.

Formation

Volcanic belts may be formed by multiple tectonic settings. They may be formed by subduction zones, which is an area on Earth where two tectonic plates meet and move towards one another, with one sliding underneath the other and moving down into the mantle, at rates typically measured in centimeters per year. An oceanic plate ordinarily slides underneath a continental plate; this often creates an orogenic zone with many volcanoes and earthquakes. In a sense, subduction zones are the opposite of divergent boundaries, areas where material rises up from the mantle and plates are moving apart. An example of a subduction-zone related volcanic belt is the Okhotsk-Chukotka Volcanic Belt in northeastern Eurasia, which is one of the largest subduction-zone related volcanic provinces in the world, stretching some 3200 km and comprising about 2 million km³ of volcanic and plutonic material.

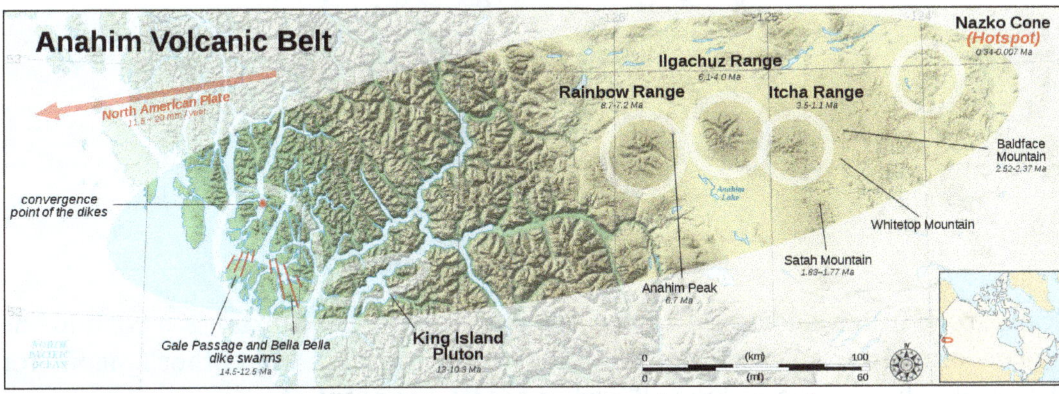

The Anahim Volcanic Belt in British Columbia, Canada.

Volcanic belts may also be formed by hotspots, which is a location on the Earth's surface that has experienced active volcanism for a long period of time. These volcanic belts are called volcanic chains. Canadian geologist John Tuzo Wilson came up with the idea in 1963 that volcanic chains like the Hawaiian Islands result from the slow movement of a tectonic plate across a "fixed" hot spot deep beneath the surface of the planet, thought to be caused by a narrow stream of hot mantle convecting up from the mantle-core boundary called a mantle plume. But more recently some geologists, such as Gillian Foulger view upper-mantle convection as a cause. This in turn has re-raised the antipodal pair impact hypothesis, the idea that pairs of opposite hot spots may result from the impact of a large meteor. Geologists have identified some 40-50 such hotspots around the globe, with Hawaii, Réunion, Yellowstone, Galápagos, and Iceland overlying the most currently active. An example of a hotspot volcanic belt is the Anahim Volcanic Belt in British Columbia, Canada, which was formed as a result of the North American Plate sliding westward over the Anahim hotspot.

Most hotspot volcanoes are basaltic because they erupt through oceanic lithosphere (e.g., Hawaii, Tahiti). As a result, they are less explosive than subduction zone volcanoes, which have high water contents. Where hotspots occur under continental crust, basaltic magma is trapped in the less dense continental crust, which is heated and melts to form rhyolites. These rhyolites can be quite hot and form violent eruptions, despite their low water content. For example, the Yellowstone Caldera was formed by some of the most powerful volcanic explosions in geologic history.

Volcanic Rock

Ignimbrite is a deposit of a pyroclastic flow.

Volcanic rock (often shortened to volcanics in scientific contexts) is a rock formed from lava erupted from a volcano. In other words, it differs from other igneous rock by being of volcanic origin. Like all rock types, the concept of volcanic rock is artificial, and in nature volcanic rocks grade into hypabyssal and metamorphic rocks and constitute an important element of some sediments and sedimentary rocks. For these reasons, in geology, volcanics and shallow hypabyssal rocks are not always treated as distinct. In the context of Precambrian shield geology, the term "volcanic" is often applied to what are strictly metavolcanic rocks. Volcanic rocks and sediment that form from magma erupted into the air are called "volcaniclastics," and these are technically sedimentary rocks.

Volcanic rocks are among the most common rock types on Earth's surface, particularly in the oceans. On land, they are very common at plate boundaries and in flood basalt provinces. It has been estimated that volcanic rocks cover about 8% of the Earth's current land surface.

Characteristics of Volcanic Rock

Setting and Size

- Lava,
- Tephra,
- Volcanic bomb,
- Lapilli,
- Volcanic ash.

Classification of Volcaniclastic rocks and sediments			
		Pyroclastic deposit	
Clast size in mm	Pyroclast	Primarily unconsolidated: tephra	Primarily consolidated: pyroclastic rock
> 64 mm	Bomb, block	Agglomerate, bed of blocks or bomb, block tephra	Agglomerate, pyroclastic breccia
64 to 2 mm	Lapillus	Layer, bed of lapilli or lapilli tephra	Lapilli tuff
2 to 1/16 mm	Coarse ash grain	Coarse ash	Coarse (ash tuff)
< 1/16 mm	Fine ash grain (dust grain)	Fine ash (dust)	Fine (ash) tuff (dust tuff)

Texture

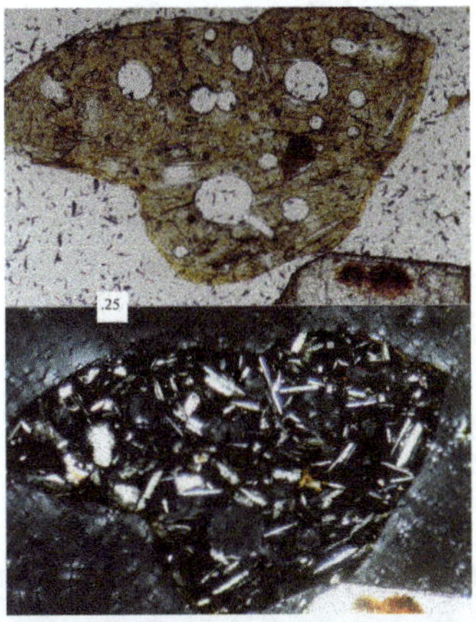

Photomicrograph of a volcanic lithic fragment (sand grain); upper picture is plane-polarized light, bottom picture is cross-polarized light, scale box at left-center is 0.25 millimeter.

Volcanic rocks are usually fine-grained or aphanitic to glass in texture. They often contain clasts of other rocks and phenocrysts. Phenocrysts are crystals that are larger than the matrix and are identifiable with the unaided eye. Rhomb porphyry is an example with large rhomb shaped phenocrysts embedded in a very fine grained matrix.

Volcanic rocks often have a vesicular texture caused by voids left by volatiles trapped in the molten lava. Pumice is a highly vesicular rock produced in explosive volcanic eruptions.

Chemistry

Most modern petrologists classify igneous rocks, including volcanic rocks, by their chemistry when dealing with their origin. The fact that different mineralogies and textures may be developed from the same initial magmas has led petrologists to rely heavily on chemistry to look at a volcanic rock's origin.

The chemistry of volcanic rocks is dependent on two things: the initial composition of the primary magma and the subsequent differentiation. Differentiation of most volcanic rocks tends to increase the silica (SiO_2) content, mainly by crystal fractionation.

The initial composition of most volcanic rocks is basaltic, albeit small differences in initial compositions may result in multiple differentiation series. The most common of these series are tholeiitic, calc-alkaline, and alkaline.

Mineralogy

Most volcanic rocks share a number of common minerals. Differentiation of volcanic rocks tends to increase the silica (SiO_2) content mainly by fractional crystallization. Thus, more evolved volcanic rocks tend to be richer in minerals with a higher amount of silica such as phyllo and tectosilicates including the feldspars, quartz polymorphs and muscovite. While still dominated by silicates, more primitive volcanic rocks have mineral assemblages with less silica, such as olivine and the pyroxenes. Bowen's reaction series correctly predicts the order of formation of the most common minerals in volcanic rocks.

Occasionally, a magma may pick up crystals that crystallized from another magma; these crystals are called xenocrysts. Diamonds found in kimberlites are rare but well-known xenocrysts; the kimberlites do not create the diamonds, but pick them up and transport them to the surface of the Earth.

Composition of Volcanic Rocks

'A'ā next to pāhoehoe lava at the Craters of the Moon National Monument and Preserve, Idaho, United States.

The sub-family of rocks that form from volcanic lava are called igneous volcanic rocks (to differentiate them from igneous rocks that form from magma below the surface, called igneous plutonic rocks).

The lavas of different volcanoes, when cooled and hardened, differ much in their appearance and composition. If a rhyolite lava-stream cools quickly, it can quickly freeze into a black glassy substance called obsidian. When filled with bubbles of gas, the same lava may form the spongy appearing pumice. Allowed to cool slowly, it forms a light-colored, uniformly solid rock called rhyolite.

The lavas, having cooled rapidly in contact with the air or water, are mostly finely crystalline or have at least fine-grained ground-mass representing that part of the viscous semi-crystalline lava flow that was still liquid at the moment of eruption. At this time they were exposed only to atmospheric pressure, and the steam and other gases, which they contained in great quantity were free to escape; many important modifications arise from this, the most striking being the frequent presence of numerous steam cavities (vesicular structure) often drawn out to elongated shapes subsequently filled up with minerals by infiltration (amygdaloidal structure).

A German example of latite, a type of volcanic rock.

As crystallization was going on while the mass was still creeping forward under the surface of the Earth, the latest formed minerals (in the ground-mass) are commonly arranged in subparallel winding lines that follow the direction of movement (fluxion or fluidal structure)—and larger early minerals that previously crystallized may show the same arrangement. Most lavas fall considerably below their original temperatures before emitted. In their behavior, they present a close analogy to hot solutions of salts in water, which, when they approach the saturation temperature, first deposit a crop of large, well-formed crystals (labile stage) and subsequently precipitate clouds of smaller less perfect crystalline particles (metastable stage).

A sample of rhyolite.

In igneous rocks the first generation of crystals generally forms before the lava has emerged to the surface, that is to say, during the ascent from the subterranean depths to the crater of the volcano. It has frequently been verified by observation that freshly emitted lavas contain large crystals borne along in a molten, liquid mass. The large, well-formed, early crystals (phenocrysts) are said to be porphyritic; the smaller crystals of the surrounding matrix or ground-mass belong to the post-effusion stage. More rarely lavas are completely fused at the moment of ejection; they may then cool to form a non-porphyritic, finely crystalline rock, or if more rapidly chilled may in large part be non-crystalline or glassy (vitreous rocks such as obsidian, tachylyte, pitchstone).

Basaltic scoria from Amsterdam Island in the Indian Ocean.

A common feature of glassy rocks is the presence of rounded bodies (spherulites), consisting of fine divergent fibres radiating from a center; they consist of imperfect crystals of feldspar, mixed with quartz or tridymite; similar bodies are often produced artificially in glasses that are allowed to cool slowly. Rarely these spherulites are hollow or consist of concentric shells with spaces between (lithophysae). Perlitic structure, also common in glasses, consists of the presence of concentric rounded cracks owing to contraction on cooling.

Volcanic rocks, Porto Moniz, Madeira.

The phenocrysts or porphyritic minerals are not only larger than those of the ground-mass; as the matrix was still liquid when they formed they were free to take perfect crystalline shapes, without interference by the pressure of adjacent crystals. They seem to have grown rapidly, as they are often filled with enclosures of glassy or finely crystalline material like that of the ground-mass . Microscopic examination of the phenocrysts often reveals that they have had a complex history. Very frequently they show layers of different composition, indicated by variations in color or other optical properties; thus augite may be green in the center surrounded by various shades of brown; or they may be pale green centrally and darker green with strong pleochroism (aegirine) at the periphery.

In the feldspars the center is usually richer in calcium than the surrounding layers, and successive zones may often be noted, each less calcic than those within it. Phenocrysts of quartz (and of other minerals), instead of sharp, perfect crystalline faces, may show rounded corroded surfaces, with the points blunted and irregular tongue-like projections of the matrix into the substance of the crystal. It is clear that after the mineral had crystallized it was partly again dissolved or corroded at some period before the matrix solidified.

Corroded phenocrysts of biotite and hornblende are very common in some lavas; they are surrounded by black rims of magnetite mixed with pale green augite. The hornblende or biotite substance has proved unstable at a certain stage of consolidation, and has been replaced by a paramorph of augite and magnetite, which may partially or completely substitute for the original crystal but still retains its characteristic outlines.

Volcanic Gas

Volcanic gases entering the atmosphere with dust and tephra during eruption of Augustine Volcano, 2006.

Volcanic gases are gases given off by active (or, at times, by dormant) volcanoes. These include gases trapped in cavities (vesicles) in volcanic rocks, dissolved or dissociated gases in magma and lava, or gases emanating directly from lava or indirectly through ground water heated by volcanic action.

Eruption of Mount St. Helens.

The sources of volcanic gases on Earth include:

- Primordial and recycled constituents from the Earth's mantle,
- Assimilated constituents from the Earth's crust,
- Groundwater and the Earth's atmosphere.

Substances that may become gaseous or give off gases when heated are termed volatile substances.

Composition

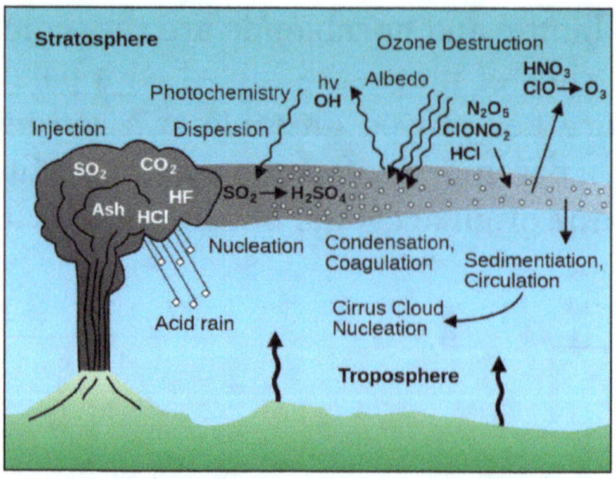

Schematic of volcanic eruption.

The principal components of volcanic gases are water vapor (H_2O), carbon dioxide (CO_2), sulfur either as sulfur dioxide (SO_2) (high-temperature volcanic gases) or hydrogen sulfide (H_2S) (low-temperature volcanic gases), nitrogen, argon, helium, neon, methane, carbon monoxide and hydrogen. Other compounds detected in volcanic gases are oxygen (meteoric), hydrogen chloride, hydrogen fluoride, hydrogen bromide, nitrogen oxide (NO_x), sulfur hexafluoride, carbonyl sulfide, and organic compounds. Exotic trace compounds include mercury, halocarbons (including CFCs), and halogen oxide radicals.

The abundance of gases varies considerably from volcano to volcano, with volcanic activity and with tectonic setting. Water vapour is consistently the most abundant volcanic gas, normally comprising more than 60% of total emissions. Carbon dioxide typically accounts for 10 to 40% of emissions.

Volcanoes located at convergent plate boundaries emit more water vapor and chlorine than volcanoes at hot spots or divergent plate boundaries. This is caused by the addition of seawater into magmas formed at subduction zones. Convergent plate boundary volcanoes also have higher H_2O/H_2, H_2O/CO_2, CO_2/He and N_2/He ratios than hot spot or divergent plate boundary volcanoes.

Magmatic Gases and High-temperature Volcanic Gases

Magma contains dissolved volatile components, as described. The solubilities of the different volatile constituents are dependent on pressure, temperature and the composition of the magma. As magma ascends towards the surface, the ambient pressure decreases, which decreases the solubility of the dissolved volatiles. Once the solubility decreases below the volatile concentration, the volatiles will tend to come out of solution within the magma (exsolve) and form a separate gas phase (the magma is super-saturated in volatiles).

The gas will initially be distributed throughout the magma as small bubbles, that cannot rise quickly through the magma. As the magma ascends the bubbles grow through a combination of expansion through decompression and growth as the solubility of volatiles in the magma decreases further causing more gas to exsolve. Depending on the viscosity of the magma, the bubbles may start to rise through the magma and coalesce, or they remain relatively fixed in place until they begin to connect and form a continuously connected network. In the former case, the bubbles may rise through the magma and accumulate at a vertical surface, e.g. the 'roof' of a magma chamber. In volcanoes with an open path to the surface, e.g. Stromboli in Italy, the bubbles may reach the surface and as they pop small explosions occur. In the latter case, the gas can flow rapidly through the continuous permeable network towards the surface. This mechanism has been used to explain activity at Santiaguito, Santa Maria volcano, Guatemala and Soufrière Hills Volcano, Montserrat. If the gas cannot escape fast enough from the magma, it will fragment the magma into small particles of ash. The fluidised ash has a much lower resistance to motion than the viscous magma, so accelerates, causing further expansion of the gases and acceleration of the mixture. This sequence of events drives explosive volcanism. Whether gas can escape gently (passive eruptions) or not (explosive eruptions) is determined by the total volatile contents of the initial magma and the viscosity of the magma, which is controlled by its composition.

The term 'closed system' degassing refers to the case where gas and its parent magma ascend together and in equilibrium with each other. The composition of the emitted gas is in equilibrium with the composition of the magma at the pressure, temperature where the gas leaves the system. In 'open system' degassing, the gas leaves its parent magma and rises up through the overlying magma without remaining in equilibrium with that magma. The gas released at the surface has a composition that is a mass-flow average of the magma exsolved at various depths and is not representative of the magma conditions at any one depth.

Molten rock (either magma or lava) near the atmosphere releases high-temperature volcanic gas (>400 °C). In explosive volcanic eruptions, the sudden release of gases from magma may cause rapid movements of the molten rock. When the magma encounters water, seawater, lake water or groundwater, it can be rapidly fragmented. The rapid expansion of gases is the driving mechanism of most explosive volcanic eruptions. However, a significant portion of volcanic gas release occurs during quasi-continuous quiescent phases of active volcanism.

Low-temperature Volcanic Gases and Hydrothermal Systems

As magmatic gas travelling upward encounters meteoric water in an aquifer, steam is produced. Latent magmatic heat can also cause meteoric waters to ascend as a vapour phase. Extended fluid-rock interaction of this hot mixture can leach constituents out of the cooling magmatic rock and also the country rock, causing volume changes and phase transitions, reactions and thus an increase in ionic strength of the upward percolating fluid. This process also decreases the fluid's pH. Cooling can cause phase separation and mineral deposition, accompanied by a shift toward more reducing conditions. At the surface expression of such hydrothermal systems, low-temperature volcanic gases (<400 °C) are either emanating as steam-gas mixtures or in dissolved form in hot springs. At the ocean floor, such hot supersaturated hydrothermal fluids form gigantic chimney structures called black smokers, at the point of emission into the cold seawater.

Non-explosive Volcanic Gas Release

The gas release can occur by advection through fractures, or via diffuse degassing through large areas of permeable ground as diffuse degassing structures (DDS). At sites of advective gas loss, precipitation of sulfur and rare minerals forms sulfur deposits and small sulfur chimneys, called fumaroles. Very low-temperature (below 100 °C) fumarolic structures are also known as solfataras. Sites of cold degassing of predominantly carbon dioxide are called mofettes. Hot springs on volcanoes often show a measurable amount of magmatic gas in dissolved form.

Sensing, Collection and Measurement

Volcanic gases were collected and analysed as long ago as 1790 by Scipione Breislak in Italy. The composition of volcanic gases is dependent on the movement of magma within the volcano. Therefore, sudden changes in gas composition often presage a change in volcanic activity. Accordingly, a large part of hazard monitoring of volcanoes involves regular measurement of gaseous emissions. For example, an increase in the CO_2 content of gases at Stromboli has been ascribed to injection of fresh volatile-rich magma at depth within the system.

Volcanic gases can be sensed (measured in-situ) or sampled for further analysis. Volcanic gas sensing can be:

- Within the gas by means of electrochemical sensors and flow-through infrared-spectroscopic gas cells;
- Outside the gas by ground-based or airborne remote spectroscopy e.g., Correlation spectroscopy (COSPEC), Differential Optical Absorption Spectroscopy (DOAS), or Fourier Transform Infrared Spectroscopy (FTIR).

Sulphur dioxide (SO_2) absorbs strongly in the ultraviolet wavelengths and has low background concentrations in the atmosphere. These characteristics make sulphur dioxide a good target for volcanic gas monitoring. It can be detected by satellite-based instruments, which allow for global monitoring, and by ground-based instruments such as DOAS. DOAS arrays are placed near some well-monitored volcanoes and used to estimate the flux of SO_2 emitted. The Multi-Component Gas Analyzer System (Multi-GAS) is also used to remotely measure CO_2 and SO_2. The fluxes of other gases are usually estimated by measuring the ratios of different gases within the volcanic plume, e.g. by FTIR, electrochemical sensors at the volcano crater rim, or direct sampling, and multiplying the ratio of the gas of interest to SO_2 by the SO_2 flux.

Direct sampling of volcanic gas sampling is often done by a method involving an evacuated flask with caustic solution, first used by Robert W. Bunsen (1811-1899) and later refined by the German chemist Werner F. Giggenbach (1937-1997), dubbed Giggenbach-bottle. Other methods include collection in evacuated empty containers, in flow-through glass tubes, in gas wash bottles (cryogenic scrubbers), on impregnated filter packs and on solid adsorbent tubes.

Analytical techniques for gas samples comprise gas chromatography with thermal conductivity detection (TCD), flame ionization detection (FID) and mass spectrometry (GC-MS) for gases, and various wet chemical techniques for dissolved species (e.g., acidimetric titration for dissolved CO_2, and ion chromatography for sulfate, chloride, fluoride). The trace metal, trace organic and isotopic composition is usually determined by different mass spectrometric methods.

Volcanic Gases and Volcano Monitoring

Certain constituents of volcanic gases may show very early signs of changing conditions at depth, making them a powerful tool to predict imminent unrest. Used in conjunction with monitoring data on seismicity and deformation, correlative monitoring gains great efficiency. Volcanic gas monitoring is a standard tool of any volcano observatory. Unfortunately, the most precise compositional data still require dangerous field sampling campaigns. However, remote sensing techniques have advanced tremendously through the 1990s. The Deep Earth Carbon Degassing Project is employing Multi-GAS remote sensing to monitor 9 volcanoes on a continuous basis.

Hazards

Volcanic gases were directly responsible for approximately 3% of all volcano-related deaths of humans between 1900 and 1986. Some volcanic gases kill by acidic corrosion; others kill by asphyxiation. The greenhouse gas, carbon dioxide, which is odorless, is emitted from volcanoes, accounting for nearly 1% of the annual global total. Some volcanic gases including sulfur dioxide, hydrogen chloride, hydrogen sulfide and hydrogen fluoride react with other atmospheric particles to form aerosols.

Rift Zone

A rift zone is a feature of some volcanoes, especially shield volcanoes, in which a set of linear cracks (or rifts) develops in a volcanic edifice, typically forming into two or three well-defined regions along the flanks of the vent. Believed to be primarily caused by internal and gravitational stresses generated by magma emplacement within and across various regions of the volcano, rift zones allow the intrusion of magmatic dykes into the slopes of the volcano itself. The addition of these magmatic materials usually contributes to the further rifting of the slope, in addition to generating fissure eruptions from those dykes that reach the surface. It is the grouping of these fissures, and the dykes that feed them, that serves to delineate where and whether a rift zone is to be defined. The accumulated lava of repeated eruptions from rift zones along with the endogenous growth created by magma intrusions causes these volcanoes to have an elongated shape. Perhaps the best example of this is Mauna Loa, which in Hawaiian means "long mountain", and which features two very well defined rift zones extending tens of kilometers outward from the central vent.

East Rift Zone on Kīlauea, Hawaii.

Formation

Rift zones are characterized by the close grouping of intrusive dykes and extrusive fissures extending outward along a relatively narrow band from the area of a central vent. The internal extensional forces and isostatic loading generated by intruding magma volumes (either associated with the magma chamber or subsequent dyke and sill formation extending outward from that chamber), in conjunction with accumulation of erupted materials, contribute to the mass and slope of the forming edifice. It is the weight of the edifice exceeding its material strength, with the additional stresses of the magma inflating the internal regions of the edifice, that can generate the initial cracking around a developing volcanic summit. Additionally, tectonic activity such as normal faulting is also commonly associated with formation of rifts along volcanic flanks. Following the path of least resistance, subsequent magmatic dykes form along and within these initial cracks, causing additional stresses to be imparted to the local materials of the edifice, which in turn generate new rifts for the magma to flow towards. In this way, established rift zones can potentially be self-sustaining geologic features along the flanks of the given volcanic vent. The orientation of this rifting is largely dependent on the gravitational and tectonic stresses at play. Basaltic shield volcanoes typically feature two main rift zones, situated with angles of 120° between in ideal situations. On shield volcanoes forming from level seafloor without neighboring vents, flank rifting occurs more evenly distributed around the vent. However, where the flanks of a volcano may be supported on one side by the presence of a pre-existing feature, or burdened with various planes of weakness, rift zone formation promulgates according to down-slope pull of gravity.

Structure

The infill of magmas in the form of dykes helps to define the shape of a volcano. A higher frequency of intrusive events along rift zones leads to elongated topographies of the affected edifices. Mathematical models show how the presence of rift zones contributes to a central horizontal bulge or ridge parallel to the orientation of the rifts. This same modelling shows how this central bulge is dependent on the ration between rift zone length and depth of the magma sources, with longer fissures over shallower sources being more positively associated with very elongated topographies of the associated flanks. Occasionally, fissure eruptions associated with rift zones can actually evolve into new vents along the volcanic edifice, generating lava flows lasting for months or longer. These lava flows add surface materials to the slopes of the volcano, extending the slopes outward in a general flattening of the morphology of the flank. The extensional character of these events can contribute to flank instability and mass wasting events where whole sections of the volcanic edifice can collapse along rift zone boundaries. These mass wasting events can affect the dyke formations and orientations as the mass of the edifice shifts, which can have profound impacts on the structural development of the edifice, while also potentially creating many volcanic hazards, such as tsunamis and dramatic shifts in directions of lava flows, to unsuspecting communities.

Volcanologist George P.L. Walker stated that rift zones were common in most volcanoes around the world, regardless of their type and formation. Walker put forward the idea that, absent any obvious signs of rifting on the surface, the presence of other volcanic features that are also associated with dyke intrusions (such as elongated cinder cones and linearly-aligned fissure vents) should also be taken to represent the presence of a rift zone-like processes in the given region. Therefore, rift zones of various lengths and widths can be tentatively identified on many stratovolcanoes and monogenetic lava fields in addition to classic Hawaiian shield volcanoes.

Examples of Rift Zone

- Hawaii: Most Hawaiian volcanoes have two or sometimes three rift zones:
    - Māhukona,
    - Mauna Loa,
    - Hualalai,
    - Kilauea,

- Galapagos Islands,
- Canary Islands:
    - La Palma,
    - El Hierro.
- Newberry Volcano, Oregon.

Magma Supply Rate

The magma supply rate measures the production rate of magma at a volcano. Global magma production rates on Earth are about 20–25 cubic kilometres per year (4.8–6.0 cu mi/a).

Magma supply rate is also known as the Armstrong unit, where 1 Armstrong Unit = 1 cubic kilometre per year (32 m^3/s). Armstrong unit can also refer to volcanic flux rate per length of arc in discussions of volcanic arcs, in that case km^2/year.

Sometimes in discussion of large volcanic systems such as volcanic arcs the volcanic flux rate is normalized to a surface area, similar to Darcy's law in hydrodynamics. It is often easier to measure magma supply rates when they are normalized for an exposed surface area as it is often difficult to delimit an intrusion.

Measurement Difficulties

Estimating the volcanic flux rate or magma supply of a volcanic system is inherently difficult for a number of reasons, and different measurements can come to different conclusions about the volcanic flux rate of a given volcanic system. Not all volcanic bodies are equally well exposed, and it is often impossible or difficult to measure magma supply rates exactly. Furthermore, volcanic flux rates often vary over time, with distinct lulls and pulses. Wall rocks may be assimilated by magma or magma may undergo differentiation such as crystallization. Magma contains vesicles and volcanic edifices are often eroded. The sizes of volcanic edifices and plutons are difficult to estimate, especially in intrusions which are mostly buried.

Applications

The magma supply rate is used to infer the behaviour of volcanic systems which erupt periodically, as well as to describe the growth of the continental crust and of deep-seated magmatic bodies such as plutons. Magma output is usually larger in oceanic settings than in continental ones, and basaltic volcanic systems produce more magma than silicic ones.

Magma Chamber

A magma chamber is a large pool of liquid rock beneath the surface of the Earth. The molten rock, or magma, in such a chamber is under great pressure, and, given enough time, that pressure can gradually fracture the rock around it, creating a way for the magma to move upward. If it finds its way to the surface, then the result will be a volcanic eruption; consequently, many volcanoes are situated over magma chambers. These chambers are hard to detect deep within the Earth, and therefore most of those known are close to the surface, commonly between 1 km and 10 km down.

Dynamics of Magma Chambers

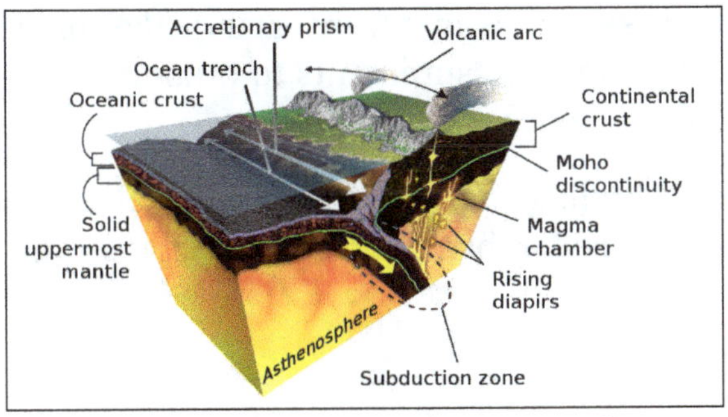

Magma chambers above a subducting plate.

Magma rises through cracks from beneath and across the crust because it is less dense than the surrounding rock. When the magma cannot find a path upwards it pools into a magma chamber. These chambers are commonly built up over time, by successive horizontal or vertical magma injections. Influx of new magma causes reaction of pre-existing crystals and the pressure in the chamber to increase.

The residing magma starts to cool, with the higher melting point components such as olivine crystallizing out of the solution, particularly near to the cooler walls of the chamber, and forming a denser conglomerate of minerals which sinks (cumulative rock). Upon cooling, new mineral phases saturate and the rock type changes (e.g. fractional crystallization), typically forming gabbro, diorite, tonalite and granite or gabbro, diorite, syenite and granite. If magma resides in a chamber for a long period, then it can become stratified with lower density components rising to the top and denser materials sinking. Rocks accumulate in layers, forming a layered intrusion. Any subsequent eruption may produce distinctly layered deposits; for example, the deposits from the 79 AD eruption of Mount Vesuvius include a thick layer of white pumice from the upper portion of the magma chamber overlaid with a similar layer of grey pumice produced from material erupted later from lower in the chamber.

Another effect of the cooling of the chamber is that the solidifying crystals will release the gas (primarily steam) previously dissolved when they were liquid, causing the pressure in the chamber to rise, possibly sufficiently to produce an eruption. Additionally, the removal of the lower melting point components will tend to make the magma more viscous (by increasing the concentration of silicates). Thus, stratification of a magma chamber may result in an increase in the amount of gas within the magma near the top of the chamber, and also make this magma more viscous, potentially leading to a more explosive eruption than would be the case had the chamber not become stratified.

If the magma is not vented to the surface in a volcanic eruption, it will slowly cool and crystallize at depth to form an intrusive igneous body, one, for example, composed of granite or gabbro.

Often, a volcano may have a deep magma chamber many kilometers down, which supplies a shallower chamber near the summit. The location of magma chambers can be mapped using seismology: seismic waves from earthquakes move more slowly through liquid rock than solid, allowing measurements to pinpoint the regions of slow movement which identify magma chambers.

As a volcano erupts, surrounding rock will collapse into the emptying chamber. If the chamber's size is reduced considerably, the resulting depression at the surface can form a caldera.

Lava Balloon

A lava balloon is a gas-filled bubble of lava that floats on the sea surface. It can be up to several metres in size. When it emerges from the sea, it is usually hot and often steaming. After floating for some time it fills with water and sinks again.

Lava balloons can form in lava flows entering the sea and at volcanic vents, but they are not common. They have been observed in the Azores, Canary Islands, Hawaii, Japan, Mariana Islands and Mexico. Apparently, they are generated when gases trapped within magma form large bubbles that eventually rise to the sea surface. In the Canary Islands, balloons containing sediments were used to infer the age of the basement on which the volcano is constructed; these sediments were also at first misinterpreted as evidence of an impending large explosive eruption.

Appearance

Lava balloons are gas-filled bubbles surrounded by a crust formed by lava; their gas content allows them to float on the sea surface. Observed sizes range from 0.3 metres (1 ft 0 in) at El Hierro (Canary Islands) during the 2011–2012 eruption to about 3 metres (9.8 ft) at Terceira on their long axis with rounded shapes. They have one or sometimes several large cavities surrounded by a 3–8-centimetre-thick (1.2–3.1 in) crust. The outer part of the crust is highly vesicular and striated and has delicate flow structures that can be seen using a scanning electron microscope. It is fragile and often breaks off the balloon. The inner part of the crust is separated from the outer part by orange and white layers. It is subdivided into three inward-thickening layers, all of which contain varying amounts of vesicles that become larger toward the interior.

A lava balloon at El Hierro.

Occurrence

Lava balloons have been described from Terceira Island in the Azores, at Teishi Knoll of Izu-Tobu (Japan) in 1989, El Hierro, offshore Pantelleria (Foerstner volcano, Italy) in 1891 and Kealakekua Bay (Mauna Loa, Hawaii) in 1877. Similar floating scoria blocks containing reticulite were observed in 1993–1994 at Socorro, Mexico. As of 2012, lava balloons have been observed only at these sites, although the increasing number of observations might indicate that this is a common mode of submarine volcanism.

A similar style of eruption but involving silicic magmas has also been found and christened "Tangaroan", after the research ship RV Tangaroa that carried out research on the Macauley caldera. Balloon-like structures were observed in 1934–1935 at Shin-Iwo-jima, Japan, and at West Rota in the Marianas. At Macauley Island in the Kermadec Islands such a style of eruption has been inferred and used to explain the presence of large rocks at substantial distances from the volcanic vent.

Observations

Lava balloons observed during a 1998–2000 eruption at Terceira are considered to be the most noteworthy expression of that eruption. They were described as steaming dark objects floating on the sea, hot enough to damage fishing ropes. At first, they were thought to be dead whales or trunks. They surfaced in batches over a span of several months, clustering in particular areas that appear to reflect the position of active volcanic vents on the seafloor but also wind and ocean current driven transport. Sometimes, hundreds of balloons were observed on a given occasion, accompanied by gas bubbles and particles shed by the balloons, all of which rose through the water in the form of plumes. The balloons steamed at first under their own heat, forming small vapour plumes and hissing sounds. Their insides could reach temperatures of over 900 °C (1,650 °F) and were sometimes incandescent. Balloons usually floated for less than 15 minutes before sinking again as water penetrated them through cracks in the crust and gases escaped. Sometimes, however, explosions threw fragments for tens of meters when water interacted with a hot interior. Remotely operated underwater vehicle (ROV) observations of the putative vent area found debris that may have come from lava balloons.

The Pantelleria eruption generated scoriaceous and vesicular floating structures with sizes exceeding 1 metre (3 ft 3 in) that sank again beneath the water surface after they had become saturated with water. 1892 descriptions of lava balloons about the Pantelleria eruption resemble the Terceira balloons. As reported by fishers, black balloons of lava floated on the sea, sometimes propelled by steam jets and sometimes exploding with up to 20 metres (66 ft) high debris fountains. As with Terceira, they were accompanied by gas bubbles and many of them were hot enough to melt zinc. Eventually, the balloons filled with water and sank again.

At El Hierro, lava balloons were erupted from 27 November 2011 until 23 February 2012 and often exploded upon reaching the sea surface. On the seafloor close to the vent were balloons with various shapes including amphora-like and sizes reaching over 4 metres (13 ft). They had sunk to the seafloor immediately after being ejected from the vent and had sometimes spilled magma. The amphora-like shape appears to have formed when floating balloons degassed through vents at their top and the balloons deformed.

Towards the end of the eruption, some lava balloons had a thin layer of solidified magma around a glassy core and appeared to float for longer times, allowing them to reach the coast. The balloons were named "restingoliths" and the glassy core "xeno-pumice". Similar balloons were observed at Teishi Knoll and appear to form when sediments are incorporated into lava and melted, forming a pumice-like structure. At El Hierro, the origin of the cores gave rise to a scientific debate about whether they originated as sediment or as silicic magma; now there is agreement that they formed out of sediments. In Socorro, the cores of lava balloons contained reticulite.

In Kealakekua Bay, over a hundred lava balloons were observed. They emitted sulfurous gases and steam and were hot inside, even incandescent. As ships were moving across the area rising balloons in the water impacted their hulls but did not do any damage.

Genesis

Large floating pumice blocks such as these observed in Kikai, Japan, in 1934–1935 may be comparable to lava balloons, but they are produced by eruptions of felsic magma, which are rich in silicates and lighter elements. By contrast, lava balloons are generally produced by eruptions of alkali basalt, although few basaltic eruptions produce them.

Lava balloons are probably limited to a depth range of 30–1,000 metres (98–3,281 ft): too deep, and gas bubbles do not form; too shallow, and degassing fragments the rocks. Only a few sufficiently large balloons can rise all the way to the sea surface; smaller ones fill quickly with water and sink. Overly crystalline magma may render a crust too brittle to form a lava balloon.

Several different mechanisms have been invoked to explain the genesis of lava balloons. Water that penetrates the lava can boil and the resulting vapours can inflate the balloons and make them float, although for Terceira a non-water gas composition has been inferred. They are usually observed when lava flows enter the sea. They appear to form when water is trapped in lava as it flows onto a beach with waves or enters lava tubes; in the latter case, entrained water can be transported through the tube and eventually end up in developing pillow lavas which are rendered buoyant by water vapour bubbles.

Less commonly, as in Terceira, balloons and accompanying gas bubbles appear to have formed on volcanic vents rather than at the front of lava flows, and more specifically on volcanic vents where magma ponded. There, gas emanating from a gas-rich magma accumulated below a crust on top of lava, forming blisters that eventually reached a critical buoyancy and broke off, forming lava balloons. The high gas content and low viscosity of the magma during the Terceira eruption allowed balloons to form despite the vents being located at considerable depth.

Finally, lava fountaining processes have been proposed to form balloons underwater. According to this model, slabs of magma in the water are surrounded by a thin shell which traps exoluting gases but also magma. The trapped gases inflate the shell and make it buoyant, while the remnant magma maintains the shell as it expands.

Impact

On São Miguel Island in the Azores, lava balloons are considered to be one of the main volcanic hazards stemming from submarine volcanic eruptions. Early lava balloons erupted during the 2011–2012 El Hierro eruption contained xeno-pumice, which raised concerns that evolved magmas such as phonolite and trachyte, capable of generating explosive eruptions, might be present under the volcano. As the eruption continued, these concerns together with an outburst of gas led to the evacuation of the town of La Restinga. The link between xeno-pumice and evolved magmas was contested early on; when explosive eruptions did not occur, this led to complaints that the response to the eruption had been disproportionate especially given its effect on the economy. The management of the El Hierro eruption in general attracted intense criticism.

Scientific Significance

At El Hierro, foraminifera fossils found in the glassy cores of lava balloons have been inferred to originate from sediments that underlie the El Hierro volcano. These fossils indicate a Cretaceous–Pliocene age for these sediments, implying that El Hierro rests on the youngest sediment base of the archipelago. The progressively lower age of the islands from east to west reinforces the theory that the Canary Islands are on top of a hotspot. Furthermore, it has been proposed that lava balloons might be proof of shallow-water volcanic eruptions.

Volcanic Hazards

A volcanic hazard is the probability that a volcanic eruption or related geophysical event will occur in a given geographic area and within a specified window of time. The risk that can be associated with a volcanic hazard depends on the proximity and vulnerability of an asset or a population of people near to where a volcanic event might occur.

Lava Flows

There are different forms of effusive lava that can provide different hazards. Pahoehoe lava is smooth and ropy while Aa lava is blocky and hard. Lava flows normally follow the topography, sinking into depressions and valleys and flowing down the volcano. Lava flows will bury roads farmlands and other forms of personal property. This lava could destroy homes, cars, and lives standing in the way. Lava flows are dangerous, however, they are slow moving and this gives people time to respond and evacuate out of immediate areas. People can mitigate this hazard by not moving to valleys or depressed areas around a volcano.

Pyroclastic Materials (Tephra) and Flow

Tephra is a generalized word for the various bits of debris that are launched out of a volcano during an eruption, regardless of their size. Pyroclastic materials are generally categorized according to size: dust measures at <1/8 mm, ash is 1/8–2 mm, cinders are 2–64 mm, and bombs and blocks are both >64 mm. There are different hazards associated with the different kinds of pyroclastic materials. Dust and ash could coat cars and homes, rendering the car unable to drive with dust accumulation in the engine. They could also layer on homes and add weight to roofs causing the house to collapse. Also, ash and dust inhaled could cause long-term respiratory issues in people inhaling the particles. Cinders are flaming pieces of ejected volcanic material which could set fire to homes and wooded areas. Bombs and blocks run the risk of hitting various objects and people within range of the volcano. Projectiles can be thrown thousands of feet in the air and can be found several miles away from the initial eruption point.

A pyroclastic flow is a fast-moving (up to 700 km/hr) extremely hot (~1000 °C) mass of air and tephra that charges down the sides of a volcano during an explosive eruption.

Mudflows, Floods, Debris Flows and Avalanches

When pyroclastic materials mix with water from a nearby stream or river, they can turn the watercourse into a fast moving mudflows. These are called lahars; when the lahar contains large material such as blocks of rock and trees, it is a volcanic debris flow. Lahars can form directly from a pyroclastic material flow flowing into a river, or could possibly form after the main eruption. The latter is referred to as secondary lahars and form when rain wets the ash and debris already on a landscape and stick together rolling along the topography. It's estimated that it can only take 30% water to initiate ash into a lahar. The thicker and/or more fast-moving a lahar, the more potential to destroy things in its path, thus making it more dangerous than a slower and/or more diluted lahar. Lahars and mudflows can damage buildings, wildlife and cars and can prove difficult to escape once caught in them. The lahars can coat objects, wash objects away and can knock objects down by their force. Lahars, debris flows and mudflows that travel into a river or stream run the potential for crowding the waterway, forcing the water to flow outward and causing a flood. The volcanic matter could also pollute the water, making it unsafe to drink.

The debris that is ejected from the volcano adds to the sides of the slope with each eruption, making the sides steeper each time. Eventually the slope gets so steep that it fails and an avalanche ensues. These avalanches carry material and debris for very long distances at very short intervals. This makes a warning system nearly immposible because the slope failure could occur at any time. The avalanche will destroy anything in its path including personal property, houses, buildings, vehicles and possibly even wildlife. If the impact of the materials in the avalanche doesn't destroy the person or object at first contact, there could be damage resulting from the weight of prolonged material on the objects.

Earthquakes Related to Volcanism

Earthquakes can occur due to volcanic activity. These earthquakes could produce topographical deformation and/or destruction of buildings, homes, cars, etc. There are two different types of these earthquakes: volcano tectonic earthquakes and long period earthquakes. "Earthquakes produced by stress changes in solid rock due to the injection or withdrawal of magma (molton rock) are called volcano tectonic earthquakes". These are hazardous due to the possibility of ground cracks or slope failures, therefore destroying everything in its path. Long period earthquakes, which happen when magma is suddenly forced into the surrounding rocks, are generally seen as a precursor to the actual eruption.

Monitoring and Mitigation

According to John Ewert and Ed Miller in a 1995 publication, "a great majority of the world's potentially active volcanoes are unmonitored". Of the historically active volcanoes in the world, less than one fourth are monitored. Only twenty-four volcanoes in the entire world are thoroughly monitored for activity. They also state that "seventy-five percent of the largest explosive eruptions since 1800 occurred at volcanoes that had no previous historical eruptions".

By monitoring the seismic and geological activity, the USGS can warn people ahead of time about impending danger. These volcanologists measure the size of an eruption in two ways: the eruption magnitude (by the volume or mass of magma erupted) and eruption intensity (by the rate of magma erupted). There are also various forms of satellites and imagery, such as satellite InSAR imagery, that monnitor the activity that isn't exposed to the naked eye.

However, the situation has somewhat changed with the International Decade for Natural Disaster Reduction and the Yokohama strategy since 1994. The Global Assessment of Risk (GAR) report is a biennial review and analysis of natural hazards published by the United Nations Office for Disaster Risk Reduction (UNISDR). The report implements the UN Hyogo Framework for Action.

Zadeh et al. (2014) provide an overview on Risks and Societal Implications of extreme natural hazards and an assessment of the global risk of volcanos and contains an appeal to found a worldwide volcanological organization comparable to the WMO . The EU has recently started major research programs dealing with risk assessment, compare:

- NOVAC: Network for Observation of Volcanic and Atmospheric Change,
- MULTIMO: Multidisciplinary Monitoring, Modelling and Forecasting of Volcanic Hazard, Explosive Eruption Risk and Decision Support for EU Populations Threatened by Volcanoes,
- ERUPT: Processes and Timescale of Magma Evolution in Volcanic Systems,
- E-RUPTIONS: A Satellite Telecommunication and Internet-Based Seismic Monitoring System for Volcanic Eruption Forecasting and Risk Management,
- EXPLORIS: Explosive Eruption Risk and Decision Support for EU Populations Threatened by Volcanoes.

Volcanism

Volcanism, also spelled vulcanism is any of various processes and phenomena associated with the surficial discharge of molten rock, pyroclastic fragments, or hot water and steam, including volcanoes, geysers, and fumaroles. Although volcanism is best known on Earth, there is evidence that it has been important in the development of the other terrestrial planets—Mercury, Venus, and Mars—as well as some natural satellites such as Earth's Moon and Jupiter's moon Io.

On Earth, volcanism occurs in several distinct geologic settings. Most of these are associated with the boundaries of the enormous rigid plates that make up the lithosphere—the crust and upper mantle. The majority of active terrestrial volcanoes (roughly 80 percent) and related phenomena occur where two lithospheric plates converge and one overrides the other, forcing it down into the mantle to be reabsorbed. Long curved chains of islands known as island arcs form at such subduction zones. Volcanoes of the explosive type make up many of the islands of a single arc or the inner row of islands of a double arc. All such islands that border the Pacific basin are built up from the seafloor, usually by the extrusion of basaltic and andesitic magmas.

A second major site of active volcanism is along the axis of the oceanic ridge system, where the plates move apart on both sides of the ridge and magma wells up from the mantle, creating new ocean floor along the trailing edges of both plates. Virtually all of this volcanic activity occurs underwater. In a few places the oceanic ridges are sufficiently elevated above the deep seafloor that they emerge from the ocean, and subaerial volcanism occurs. Iceland is the best-known example. The magmas that are erupted along the oceanic ridges are basaltic in composition.

A relatively small number of volcanoes occur within plates far from their margins. Some, as exemplified by the volcanic islands of Hawaii that lie in the interior of the Pacific Plate, are thought to occur because of plate movement over a "hot spot" from which magmas can penetrate to the surface. These magmas characteristically generate a chain of progressively older volcanoes that mark the direction of past motion of the plate over a particular hot spot. The active volcanoes of the East African Rift Valley also occur within a plate (the African Plate), but they appear to result from a different mechanism—possibly the beginning of a new region of plates moving apart.

Pyroclastic Flow

Pyroclastic flows sweep down the flanks of Mayon Volcano, Philippines.

A pyroclastic flow (also known as a pyroclastic density current or a pyroclastic cloud) is a fast-moving current of hot gas and volcanic matter (collectively known as tephra) that moves away from a volcano about 100 km/h (62 mph) on average but is capable of reaching speeds up to 700 km/h (430 mph). The gases can reach temperatures of about 1,000 °C (1,830 °F).

Pyroclastic flows are a common and devastating result of certain explosive eruptions; they normally touch the ground and hurtle downhill, or spread laterally under gravity. Their speed depends upon the density of the current, the volcanic output rate, and the gradient of the slope, and are capable for producing more lightning.

Rocks from the Bishop tuff, uncompressed with pumice (on left); compressed with fiamme (on right).

A name for pyroclastic flows which glow red in the dark is nuée ardente; this was first used to describe the disastrous 1902 eruption of Mount Pelée on Martinique.

Pyroclastic flows that contain a much higher proportion of gas to rock are known as "fully dilute pyroclastic density currents" or pyroclastic surges. The lower density sometimes allows them to flow over higher topographic features or water such as ridges, hills, rivers and seas. They may also contain steam, water and rock at less than 250 °C (482 °F); these are called "cold" compared with other flows, although the temperature is still lethally high. Cold pyroclastic surges can occur when the eruption is from a vent under a shallow lake or the sea. Fronts of some pyroclastic density currents are fully dilute; for example, during the eruption of Mount Pelée in 1902, a fully dilute current overwhelmed the city of Saint-Pierre and killed nearly 30,000 people.

A pyroclastic flow is a type of gravity current; in scientific literature they are sometimes abbreviated to PDC (pyroclastic density current).

Causes

There are several mechanisms that can produce a pyroclastic flow:

- Fountain collapse of an eruption column from a Plinian eruption (e.g. Mount Vesuvius' destruction of Herculaneum and Pompeii). In such an eruption, the material forcefully ejected from the vent heats the surrounding air and the turbulent mixture rises, through convection, for many kilometers. If the erupted jet is unable to heat the surrounding air sufficiently, convection currents will not be strong enough to carry the plume upwards and it falls, flowing down the flanks of the volcano.
- Fountain collapse of an eruption column associated with a Vulcanian eruption (e.g., Montserrat's Soufrière Hills volcano has generated many of these deadly

pyroclastic flows and surges). The gas and projectiles create a cloud that is denser than the surrounding air and becomes a pyroclastic flow.

- Frothing at the mouth of the vent during degassing of the erupted lava. This can lead to the production of a rock called ignimbrite. This occurred during the eruption of Novarupta in 1912.
- Gravitational collapse of a lava dome or spine, with subsequent avalanches and flows down a steep slope (e.g., Montserrat's Soufrière Hills volcano, which caused nineteen deaths in 1997).
- The directional blast (or jet) when part of a volcano collapses or explodes (e.g., the eruption of Mount St. Helens in May 18, 1980). As distance from the volcano increases, this rapidly transforms into a gravity-driven current.

Size and Effects

Building remnant in Francisco Leon destroyed by pyroclastic surges and flows during eruption of El Chichon volcano in Mexico 1982. Reinforcement rods in concrete bent in the direction of the flow.

Flow volumes range from a few hundred cubic meters (yards) to more than 1,000 cubic kilometres (~240 cubic miles). Larger flows can travel for hundreds of kilometres (milcs), although none on that scale has occurred for several hundred thousand years. Most pyroclastic flows are around 1 to 10 km^3 (about 1⁄4 to 2½ cubic miles) and travel for several kilometres. Flows usually consist of two parts: the basal flow hugs the ground and contains larger, coarse boulders and rock fragments, while an extremely hot ash plume lofts above it because of the turbulence between the flow and the overlying air, admixing and heating cold atmospheric air causing expansion and convection.

A scientist examines pumice blocks at the edge of a pyroclastic flow deposit from Mount St. Helens.

The kinetic energy of the moving cloud will flatten trees and buildings in its path. The hot gases and high speed make them particularly lethal, as they will incinerate living organisms instantaneously or turn them into carbonized fossils:

- The cities of Pompeii and Herculaneum, Italy, for example, were engulfed by pyroclastic surges on August 24, 79 AD with many lives lost.
- The 1902 eruption of Mount Pelée destroyed the Martinique town of St. Pierre. Despite signs of impending eruption, the government deemed St. Pierre safe due to hills and valleys between it and the volcano, but the pyroclastic flow charred almost the entirety of the city, killing all but two of its 30,000 residents.
- A pyroclastic surge killed volcanologists Harry Glicken and Katia and Maurice Krafft and 40 other people on Mount Unzen, in Japan, on June 3, 1991. The surge started as a pyroclastic flow and the more energised surge climbed a spur on which the Kraffts and the others were standing; it engulfed them, and the corpses were covered with about 5 mm (0.2 in) of ash.
- On 25 June, 1997 a pyroclastic flow travelled down Mosquito Ghaut on the Caribbean island of Montserrat. A large, highly energized pyroclastic surge developed. This flow could not be restrained by the Ghaut and spilled out of it, killing 19 people who were in the Streatham village area (which was officially evacuated). Several others in the area suffered severe burns.

The casts of some victims in the so-called “Garden of the Fugitives”, Pompeii.

Interaction with Water

Testimonial evidence from the 1883 eruption of Krakatoa, supported by experimental evidence, shows that pyroclastic flows can cross significant bodies of water. However, that might be a pyroclastic surge, not flow, because the density of a gravity current means it cannot move across the surface of water. One flow reached the Sumatran coast as much as 48 km (30 mi) away.

A 2006 BBC documentary film, 'Ten things you didn't know about Volcanoes', demonstrated tests by a research team at Kiel University, Germany, of pyroclastic flows moving over water. When the reconstructed pyroclastic flow (stream of mostly hot ash with varying densities) hit the water, two things happened: the heavier material fell into the water, precipitating out from the pyroclastic flow and into the liquid; the temperature of the ash caused the water to evaporate, propelling the pyroclastic flow (now only consisting of the lighter material) along on a bed of steam at an even faster pace than before.

During some phases of the Soufriere Hills volcano on Montserrat, pyroclastic flows were filmed about 1 km (0.6 mi) offshore. These show the water boiling as the flow passed over it. The flows eventually built a delta, which covered about 1 km2 (250 acres).

A pyroclastic flow can interact with a body of water to form a large amount of mud, which can then continue to flow downhill as a lahar. This is one of several mechanisms that can create a lahar.

Prediction of Volcanic Activity

Prediction of volcanic eruption (volcanic eruption forecasting) is an interdisciplinary monitoring and research effort to predict the time and severity of a volcano's eruption. Of particular importance is the prediction of hazardous eruptions that could lead to catastrophic loss of life, property, and disruption of human activities.

Mount St. Helens erupted explosively.

Seismic Waves (Seismicity)

General Principles of Volcano Seismology

- Seismic activity (earthquakes and tremors) always occurs as volcanoes awaken and prepare to erupt and are a very important link to eruptions. Some volcanoes normally have continuing low-level seismic activity, but an increase may signal a greater likelihood of an eruption. The types of earthquakes that occur and where they start and end are also key signs. Volcanic seismicity has three major forms: short-period earthquake, long-period earthquake, and harmonic tremor.

- Short-period earthquakes are like normal fault-generated earthquakes. They are caused by the fracturing of brittle rock as magma forces its way upward. These short-period earthquakes signify the growth of a magma body near the surface and are known as 'A' waves. These type of seismic events are often also referred to as Volcano-Tectonic (or VT) events or earthquakes.
- Long-period earthquakes are believed to indicate increased gas pressure in a volcano's plumbing system. They are similar to the clanging sometimes heard in a house's plumbing system, which is known as "water hammer". These oscillations are the equivalent of acoustic vibrations in a chamber, in the context of magma chambers within the volcanic dome and are known as 'B' waves. These are also known as resonance waves and long period resonance events.
- Harmonic tremors are often the result of magma pushing against the overlying rock below the surface. They can sometimes be strong enough to be felt as humming or buzzing by people and animals, hence the name.

Patterns of seismicity are complex and often difficult to interpret; however, increasing seismic activity is a good indicator of increasing eruption risk, especially if long-period events become dominant and episodes of harmonic tremor appear.

Using a similar method, researchers can detect volcanic eruptions by monitoring infra-sound—sub-audible sound below 20 Hz. The IMS Global Infrasound Network, originally set up to verify compliance with nuclear test ban treaties, has 60 stations around the world that work to detect and locate erupting volcanoes.

Iceberg Tremors

Similarities between iceberg tremors, which occur when they run aground, and volcanic tremors may help experts develop a better method for predicting volcanic eruptions. Although icebergs have much simpler structures than volcanoes, they are physically easier to work with. The similarities between volcanic and iceberg tremors include long durations and amplitudes, as well as common shifts in frequencies.

Gas Emissions

As magma nears the surface and its pressure decreases, gases escape. This process is much like what happens when you open a bottle of fizzy drink and carbon dioxide escapes. Sulfur dioxide is one of the main components of volcanic gases, and increasing amounts of it herald the arrival of increasing amounts of magma near the surface. For example, on May 13, 1991, an increasing amount of sulfur dioxide was released from Mount Pinatubo in the Philippines. On May 28, just two weeks later, sulfur dioxide emissions had increased to 5,000 tonnes, ten times the earlier amount. Mount Pinatubo later erupted on June 12, 1991. On several occasions, such as before the Mount Pinatubo eruption and the 1993 Galeras, Colombia eruption, sulfur dioxide emissions have dropped to low levels prior to eruptions. Most scientists believe that this drop in gas levels is caused by the sealing of gas passages by hardened magma. Such an event leads to increased pressure in the volcano's plumbing system and an increased chance of an explosive eruption. A multi-component gas analyzer system (Multi-GAS) is an instrument package used to take real-time high-resolution measurements of volcanic gas plumes. Multi-GAS measurements of CO_2/SO_2 ratios can allow detection of the pre-eruptive degassing of rising magmas, improving prediction of volcanic activity.

Gas and ash plume erupted from Mount Pinatubo, Philippines.

Ground Deformation

Swelling of a volcano signals that magma has accumulated near the surface. Scientists monitoring an active volcano will often measure the tilt of the slope and track changes in the rate of swelling. An increased rate of swelling, especially if accompanied by an increase in sulfur dioxide emissions and harmonic tremors is a high probability sign of an impending event. The deformation of Mount St. Helens prior to the May 18, 1980 eruption was a classic example of deformation, as the north side of the volcano was bulging upwards as magma was building up underneath. Most cases of ground deformation are usually detectable only by sophisticated equipment used by scientists, but they can still predict future eruptions this way. The Hawaiian Volcanoes show significant ground deformation; there is inflation of the ground prior to an eruption and then an obvious deflation post-eruption. This is due to the shallow magma chamber of the Hawaiian Volcanoes; movement of the magma is easily noticed on the ground above.

Thermal Monitoring

Both magma movement, changes in gas release and hydrothermal activity can lead to thermal emissivity changes at the volcano's surface. These can be measured using several techniques:

- Forward looking infrared radiometry (FLIR) from hand-held devices installed on-site, at a distance, or airborne;
- Infrared band satellite imagery;
- In-Situ thermometry (hot springs, fumaroles);
- Heat flux maps;
- Geothermal well enthalpy changes.

Hydrology

There are 4 main methods that can be used to predict a volcanic eruption through the use of hydrology:

- Borehole and well hydrologic and hydraulic measurements are increasingly used to monitor changes in a volcanoes subsurface gas pressure and thermal regime. Increased gas pressure will make water levels rise and suddenly drop right before an eruption, and thermal focusing (increased local heat flow) can reduce or dry out aquifers.
- Detection of lahars and other debris flows close to their sources. USGS scientists have developed an inexpensive, durable, portable and easily installed system to detect and continuously monitor the arrival and passage of debris flows and floods in river valleys that drain active volcanoes.
- Pre-eruption sediment may be picked up by a river channel surrounding the volcano that shows that the actual eruption may be imminent. Most sediment is transported from volcanically disturbed watersheds during periods of heavy rainfall. This can be an indication of morphological changes and increased hydrothermal activity in absence of instrumental monitoring techniques.
- Volcanic deposit that may be placed on a river bank can easily be eroded which will dramatically widen or deepen the river channel. Therefore, monitoring of the river channels width and depth can be used to assess the likelihood of a future volcanic eruption.

Remote Sensing

Remote sensing is the detection by a satellite's sensors of electromagnetic energy that is absorbed, reflected, radiated or scattered from the surface of a volcano or from its erupted material in an eruption cloud.

- Cloud sensing: Scientists can monitor the unusually cold eruption clouds from volcanoes using data from two different thermal wavelengths to enhance the visibility of eruption clouds and discriminate them from meteorological clouds.
- Gas sensing: Sulfur dioxide can also be measured by remote sensing at some of the same wavelengths as ozone. Total Ozone Mapping Spectrometers (TOMS) can measure the amount of sulfur dioxide gas released by volcanoes in eruptions. Carbon dioxide emissions from volcanoes have been detected in the short-wave infrared using NASA's Orbiting Carbon Observatory 2.
- Thermal sensing: The presence of new significant thermal signatures or 'hot spots' may indicate new heating of the ground before an eruption, represent an eruption in progress or the presence of a very recent volcanic deposit, including lava flows or pyroclastic flows.
- Deformation sensing: Satellite-borne spatial radar data can be used to detect long-term geometric changes in the volcanic edifice, such as uplift and depression. In this method, interferometric synthetic aperture radar (InSAR), digital elevation models generated from radar imagery are subtracted from each other to yield a differential image, displaying rates of topographic change.
- Forest monitoring: In recent period it has been demonstrated the location of eruptive fractures could be predicted, months to years before the eruptions, by the monitoring of forest growth. This tool based on the monitoring of the trees growth has been validated at both Mt. Niyragongo and Mt. Etna during the 2002–2003 volcano eruptive events.

Mass Movements and Mass Failures

Monitoring mass movements and -failures uses techniques lending from seismology (geophones), deformation, and meteorology. Landslides, rock falls, pyroclastic flows, and mud flows (lahars) are example of mass failures of volcanic material before, during, and after eruptions.

The most famous volcanic landslide was probably the failure of a bulge that built up from intruding magma before the Mt. St. Helens eruption in 1980, this landslide "uncorked" the shallow magmatic intrusion causing catastrophic failure and an unexpected lateral eruption blast. Rock falls often occur during periods of increased deformation and can be a sign of increased activity in absence of instrumental monitoring. Mud flows (lahars) are remobilized hydrated ash deposits from pyroclastic flows and ash fall deposits, moving downslope even at very shallow angles at high speed. Because of their high density they are capable of moving large objects such as loaded logging trucks, houses, bridges, and boulders. Their deposits usually form a second ring of debris fans around volcanic edifices, the inner fan being primary ash deposits. Downstream of the deposition of their finest load, lahars can still pose a sheet flood hazard from the residual water.

Lahar deposits can take many months to dry out, until they can be walked on. The hazards derived from lahar activity can exist several years after a large explosive eruption.

A team of US scientists developed a method of predicting lahars. Their method was developed by analyzing rocks on Mt. Rainier in Washington. The warning system depends on noting the differences between fresh rocks and older ones. Fresh rocks are poor conductors of electricity and become hydrothermically altered by water and heat. Therefore, if they know the age of the rocks, and therefore the strength of them, they can predict the pathways of a lahar. A system of Acoustic Flow Monitors (AFM) has also been emplaced on Mount Rainier to analyze ground tremors that could result in a lahar, providing an earlier warning.

Dispersal Index

Dispersal index is a parameter in volcanology. The dispersal index *D* was defined by George P. L. Walker in 1973 as the surface area covered by an ash or tephra fall, where the thickness is equal or more than 1/100 of the thickness of the fall at the vent. An eruption with a low dispersal index leaves most of its products close to the vent, forming a cone; an eruption with a high dispersal index forms thinner sheet-like deposits which extends to larger distances from the vent. A dispersal index of 500 square kilometres (190 sq mi) or more of coarse pumice is one proposed definition of a Plinian eruption. Likewise, a dispersal index of 50,000 square kilometres (19,000 sq mi) has been proposed as a cutoff for an ultraplinian eruption. The definition of 1/100 of the near-vent thickness was partially dictated by the fact that most tephra deposits are not well preserved at larger distances.

Originally, the dispersal index was considered a function of the height of the eruption column. Later, a role for the size of the tephra and ash particles was identified, with coarser fall deposits covering smaller surfaces than finer deposits generated by a column of the same height. For example, a deposit with a dispersal index of 500 square kilometres (190 sq mi) can be formed by a column with heights of 14–18 square kilometres (5.4–6.9 sq mi). Thus, Walker's idea of the column height alone separating a cone forming eruption and an eruption generating a sheet-like deposit was later considered oversimplified. An additional complicating factor is that fine particles are prone to aggregating and thus falling out more quickly from the column. Further problems arise when the maximum thickness has to be determined.

The height of the eruption column, the presence and behaviour of water, the speed and direction of the wind as well as the sizes of the various tephra particles influence the fallout patterns of an ash cloud.

The dispersal index for volcanic eruptions ranges from <1 square kilometre (0.39 sq mi) and 1–1,000 square kilometres (0.39–386.10 sq mi). A number of basaltic phreatomagmatic deposits, frequently associated with tuff rings, have a dispersal index of less than 50 square kilometres (19 sq mi).

Volcano	Eruption	Age	Dispersal index
Taupo	Hatepe eruption	1820 BP	100,000 square kilometres (39,000 sq mi)
Taupo	Oruanui eruption	~20000 BP	>100,000 square kilometres (39,000 sq mi)
Taupo	Hinemaiaia tephra	4500 years ago	40,000 square kilometres (15,000 sq mi)
Kelut		1990	2,000 square kilometres (770 sq mi)
Rinjani	1257 Samalas eruption, P1 phase	1257	7,500 square kilometres (2,900 sq mi)
Rinjani	1257 Samalas eruption, P3 phase	1257	110,500 square kilometres (42,700 sq mi)
Mount Pelée	P1 eruption	650 BP	900 square kilometres (350 sq mi)
Mount Pelée	P2 eruption	1670 BP	800 square kilometres (310 sq mi)
Mount Pelée	P3 eruption	2010 BP	1,000 square kilometres (390 sq mi)
Rabaul	Vulcan	1937	40 square kilometres (15 sq mi)
Okataina Volcanic Complex	Whakatane tephra	~ 5500 BP	~200,000 square kilometres (77,000 sq mi)
Agua de Pau	Fogo A	5000 BP	1,500 square kilometres (580 sq mi)
Hekla		1991	460 square kilometres (180 sq mi)
Sakurajima	Taisho	1914	539 square kilometres (208 sq mi)
Mono Craters		4th century AD	1,800 square kilometres (690 sq mi)

A related measure is the thickness half-distance b_t, which defines the distance over which the thickness of a deposit halves. These values are related with each other over $D = \frac{\pi b_t^2 ln(0.01)^2}{ln(0.5)^2}$ for circular deposits.

Volcanic Explosivity Index

The Volcanic Explosivity Index (VEI) is a relative measure of the explosiveness of volcanic eruptions. It was devised by Chris Newhall of the United States Geological Survey and Stephen Self at the University of Hawaii in 1982.

Volume of products, eruption cloud height, and qualitative observations (using terms ranging from "gentle" to "mega-colossal") are used to determine the explosivity value. The scale is open-ended with the largest volcanoes in history given magnitude 8. A value of 0 is given for non-explosive eruptions, defined as less than 10,000 m^3 (350,000 cu ft) of tephra ejected; and 8 representing a mega-colossal explosive eruption that can eject 1.0×10^{12} m^3 (240 cubic miles) of tephra and have a cloud column height of over 20 km (66,000 ft). The scale is logarithmic, with each interval on the scale representing a tenfold increase in observed ejecta criteria, with the exception of between VEI-0, VEI-1 and VEI-2.

Classification

With indices running from 0 to 8, the VEI associated with an eruption is dependent on how much volcanic material is thrown out, to what height, and how long the eruption lasts. The scale is logarithmic from VEI-2 and up; an increase of 1 index indicates an eruption that is 10 times as powerful. As such there is a discontinuity in the definition of the VEI between indices 1 and 2. The lower border of the volume of ejecta jumps by a factor of one hundred, from 10,000 to 1,000,000 m^3 (350,000 to 35,310,000 cu ft), while the factor is ten between all higher indices.

About 40 eruptions of VEI-8 magnitude within the last 132 million years (Mya) have been identified, of which 30 occurred in the past 36 million years. Considering the estimated frequency is on the order of once in 50,000 years, there are likely many such eruptions in the last 132 Mya that are not yet known. Based on incomplete statistics, other authors assume that at least 60 VEI-8 eruptions have been identified. The most recent is Lake Taupo's Oruanui eruption, more than 27,000 years ago, which means that there have not been any Holocene eruptions with a VEI of 8.

There have been at least 10 eruptions of VEI-7 in the last 10,000 years. There are also 58 plinian eruptions, and 13 caldera-forming eruptions, of large, but unknown magnitudes. By 2010, the Global Volcanism Program of the Smithsonian Institution had catalogued the assignment of a VEI for 7,742 volcanic eruptions that occurred during the Holocene (the last 11,700 years) which account for about 75% of the total known eruptions during the Holocene. Of these 7,742 eruptions, about 49% have a VEI of ≤ 2, and 90% have a VEI ≤ 3.

Limitations

Under the VEI, ash, lava, lava bombs, and ignimbrite are all treated alike. Density and vesicularity (gas bubbling) of the volcanic products in question is not taken into account. In contrast, the DRE (dense-rock equivalent) is sometimes calculated to give the actual amount of magma erupted. Another weakness of the VEI is that it does not take into account the power output of an eruption, which makes the VEI extremely difficult to determine with prehistoric or unobserved eruptions.

Although VEI is quite suitable for classifying the explosive magnitude of eruptions, the index is not as significant as sulphur dioxide emissions in quantifying their atmospheric and climatic impact, as a 2004 paper by Georgina Miles, Roy Grainger and Eleanor Highwood points out.

"Tephra, or fallout sediment analysis, can provide an estimate of the explosiveness of a known eruption event. It is, however, not obviously related to the amount of SO_2 emitted by the eruption. The Volcanic Explosivity Index (VEI) was derived to catalogue the explosive magnitude of historical eruptions, based on the order of magnitude of erupted mass, and gives a general indication as to the height the eruptive column reached. The VEI itself is inadequate for describing the atmospheric effects of volcanic eruptions. This is clearly demonstrated by two eruptions, Agung (1963) and El Chichón (1982). Their VEI classification separates them by an order of magnitude in explosivity, although the volume of SO_2 released into the stratosphere by each was measured to be broadly similar, as shown by the optical depth data for the two eruptions."

Monogenetic Volcanic Field

A monogenetic volcanic field is a type of volcanic field consisting of a group of small monogenetic volcanoes, each of which erupts only once, as opposed to polygenetic volcanoes, which erupt repeatedly over a period of time. Many monogenetic volcanoes are cinder cones, often with lava flows, such as Parícutin in the Michoacán-Guanajuato volcanic field, which erupted from 1943 to 1952. Some monogenetic volcanoes are small lava shields, such as Rangitoto Island in the Auckland volcanic field. Other monogenetic volcanoes are tuff rings or maars. A monogenetic field typically contains between ten and a hundred volcanoes. The Michoacán-Guanajuato field in Mexico contains more than a thousand volcanoes and is much larger than usual.

Monogenetic fields occur only where the magma supply to the volcano is low or where vents are not close enough or large enough to develop plumbing systems for continuous feeding of magma. Monogenetic volcanic fields can provide snapshots of the underlying region beneath the surface, and may be useful in studying the generation of magma and the composition of the mantle since the single eruption produced would match that of the chamber from which it erupted.

Examples of monogenetic volcanic fields:

- Auckland volcanic field, underlying much of the city of Auckland, New Zealand.
- Wells Gray-Clearwater volcanic field, east-central British Columbia, Canada.
- Honolulu Volcanic Series, Hawaii, United States.
- Boring Lava Field, in and near Portland, Oregon, United States.
- Michoacán-Guanajuato volcanic field (includes El Jorullo and Parícutin), Michoacán, Mexico.
- Vulkan Eifel, Germany.
- Chaîne des Puys, France.
- Carrán-Los Venados, Chile.
- North and north-eastern area of Gran Canaria, Canary Islands, Spain.
- Southern Volcanic Zone of Tenerife, Canary Islands, Spain.
- Newer Volcanics Province, South-eastern Australia.

Volcanic Plumbing Systems

Magma transport and storage beneath active volcanoes occurs in the so-called volcanic plumbing system (VPS), a network of different magmatic sheet intrusions and magma reservoirs. The complex physical and chemical processes, which occur in the volcanic plumbing system, are key parameters that control the occurrence of an eruption, as well as type and size of the eruption. It is therefore imperative to assess plumbing system processes and their dynamics. Traditionally, plumbing system research is done as a part of various scientific disciplines, each with its own research questions, methods, and terms.

Volcanic plumbing systems (VPS) form a plexus of magma channels and reservoirs that are governed by complex interactions of chemical and mechanical processes that control how magmas are emplaced and how they propagate through the Earth's crust to an eventual eruption. Volcanic plumbing systems thus set the stage for volcanic eruptions and govern the style and magnitude of eruptive activity including dramatic volcano–tectonic phenomena, such as caldera and sector collapses.

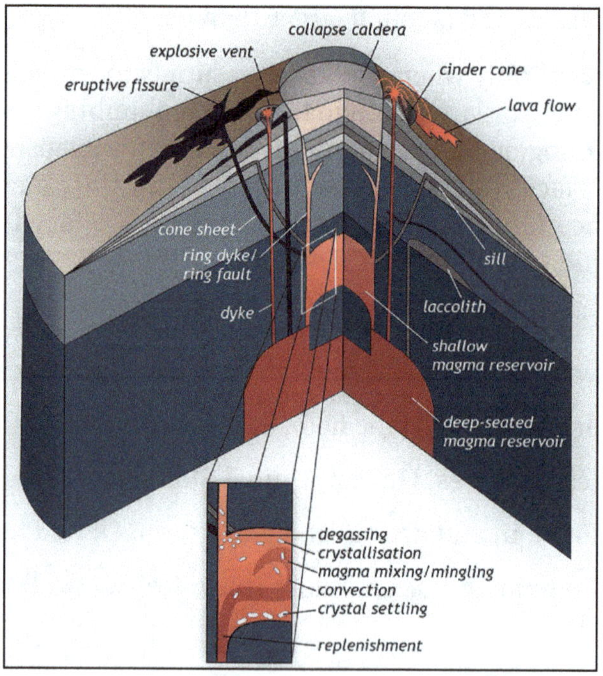

In figure, Schematic sketch of different components of volcanic plumbing systems, highlighting the complexity of the different types of magma channels and reservoirs. The inset illustrates the processes that may occur within magma bodies.

Traditionally, however, the study of the plumbing system components, such as dykes, sills, and larger magma bodies, as well as their dynamics is strongly method-based, for example, focussing exclusively on the composition of plutonic bodies or the seismicity of magma ascent. To date, relatively few bridges between the distinct disciplines exist.

Methods Commonly used to Study Volcanic Plumbing Systems

Field based studies of volcanic plumbing systems have initially aimed at a qualitative description of intrusive phenomena and focused on a classification of intrusion lithologies and morphologies to understand how magma is transported, stored, and evolves in the crust. With time, field work has become increasingly quantitative, producing a detailed record of the compositions and structures associated with magmatic intrusions. Analyses of the composition, absolute and relative ages, and dimensions of the components of magmatic plumbing systems have produced a more and more systematic view of the emplacement and evolution of plumbing systems. Besides the recording of variations in lithology and emplacement-related structures, mapping of magmatic intrusions often includes the study of magmatic fabrics recorded within the igneous rocks, such as preferred orientation of phenocrysts and of magnetic minerals using their so-called anisotropy of magnetic susceptibility (AMS;). Recently, classical methods, such as field mapping with paper maps and compass, have become complemented by modern digital mapping techniques using global positioning system (GPS) and smart phones. At the same time, the possibilities to analyse structural data collected in the field become more and more sophisticated. In addition to thorough statistical analysis and stereographic projection, three-dimensional (3D) structural modelling is used to visualise, reconstruct, and interpret structural field data.

In figure, examples of field and structural studies of volcanic plumbing systems. (a) Geological map of the Isle of Rum draped over a Digital Elevation Model. (b) High-precision

GPS mapping of the roof of a granitic pluton. (c) Result of 3D reconstruction of a granitic pluton based on GPS mapping. (d) Measuring fractures in a granitic intrusion using a smart phone and Field Move Clino app by Midland Valley Ltd. (e) Outcrop of basaltic inclined sheets and corresponding projection of poles to orientation planes and density of 321 such sheet intrusions in an equal area, lower hemisphere plot. (f) Probability Density Function (PDF) of the thickness of magmatic sheets, such as shown in (e) compared to a selection of statistical distributions. (g) 3D structural model of the plumbing system of the Ardnamurchan central complex, Scotland, produced using Move by Midland Valley Ltd.

Igneous plumbing systems in outcrops regularly extend over several kilometres. Such wide extents represent a substantial challenge for having a correct overview understanding of the structure of the exposed VPS and completing structural field surveying in a manageable time. In addition, numerous field areas of interest are hardly reachable for direct observations, such as very steep mountains and crevasses. Recent technologies, including LiDAR scanners, drones, and robots, have recently started overcoming these challenges. The one hand, LiDAR scanners have been used to produce high-precision and high-resolution textured virtual outcrop models. Such digital models allow detailed, quantitative 2D and 3D fracture mapping of extensive, mostly subvertical outcrops. On the other hand, drone surveys combined with photogrammetric tools are very helpful to produce virtual outcrop models and orthorectified images of extensive subhorizontal or gently dipping outcrops, and provide new observational perspectives for extensive structural mapping. Finally, robots equipped with monitoring tools can explore the Earth's interior that is inaccessible for humans. The data produced by these modern tools allow new possibilities for post-field digital mapping of extensive areas, thus shortening the field campaigns.

Therefore, field studies of exhumed plumbing systems form the foundation of our conceptual understanding of the individual components of volcanic plumbing systems, their morphologies, sizes, and emplacement mechanisms, as well as of characteristic structures in the host rock associated with the evolution of magmatic intrusions. Field work in eroded volcanic areas is therefore of critical importance, because it provides us with fundamental information that may be used to benchmark numerical and laboratory models and to interpret geophysical and geodetic data.

The main limitation of the field-based approach to study volcanic plumbing systems is that fossil and eroded volcanic plumbing systems represent a snapshot of the final state of the magmatic system only, while active volcanoes do not permit a detailed look inside their VPS. An outcrop in an eroded volcano is essentially a snapshot of the sum of all superimposed processes a suite of rock has experienced. It is thus not always straightforward to extract accurate age relationship between individual units or even distinct boundaries at times, as well as to deduce what dynamic processes were contributing to the final picture.

Petrological and Geochemical Studies of Volcanic Plumbing Systems

Igneous petrology and geochemistry are among the classic approaches used to characterise volcanic plumbing systems and aim to describe the conditions, time scales, and characteristics of the chemical evolution of magma. This characterisation is generally based on the mineralogy and textures of igneous rocks, as well as their major and trace element composition. In order to study the minerals, textures, and compositions of rock samples collected in the field, igneous petrologists employ a wide range of analytical techniques, which have been developed simultaneously with, and strongly facilitated, an increasing understanding of the chemical evolution of minerals and melts. These analytical techniques have emerged in a rapid succession and have become increasingly precise since igneous petrology was established as a discipline and the first microscopes were built in the second half of the nineteenth century. Today, rock textures and the type and associations of minerals in thin sections of rock samples are studied using a variety of microscopes. The bulk, or whole-rock, composition of an igneous rock that mostly comprises the quantity of major elements or their oxides can be derived through, for example, exposing a powdered sample of a rock to X-rays (X-ray fluorescence (XRF)). Other analytical techniques, such as mass spectrometry, can quantify the isotopic composition of the major and trace elements of a sample.

Data on the mineralogy and composition of igneous rocks are probes into the chemical evolution of volcanic plumbing systems and can be used in many ways. In many cases, the whole-rock composition is characteristic of the geodynamic setting and origin of a magma and can also be used to discriminate processes in the plumbing system, such as fractional crystallisation. The concentration of trace elements can be used in geochemical modelling to quantify processes in the volcanic plumbing system, such as assimilation of country rocks into the magma. The decay of radiogenic isotopes is used as a standard tool to determine the absolute age of a rock sample, which has led to detailed insights into the time scales of magma emplacement in volcanic plumbing systems.

At the scale of individual crystals, crystal size distributions, chemical zoning, and textures can serve as records of the chemical and thermal evolution of their host magmas. Crystal growth rates and chemical diffusion across crystal zones can furthermore be used to quantify the time scales of, for example, magma storage or replenishment. Moreover, pressure and/or temperature dependent mineral compositions can be used as so-called geobarometers and/or geothermometers that reveal the depth and conditions at which certain minerals grew, which usually corresponds to magma reservoir depths.

In order to quantify the chemical evolution of minerals and magmas in situ, petrological experiments are used to simulate, for example, the influence of pressure and/or temperature on mineral compositions and the reaction of magma with crustal rocks.

Petrological and geochemical studies of volcanic plumbing systems are thus the foundation of our understanding of the chemical and thermal processes during magma storage and offer insight into the time scales of magma transport and evolution. Estimates of the depth of magma storage provide valuable constraints on the interpretation of geodetic and geophysical monitoring data of active volcanoes. Furthermore, a characterisation of the types of magma erupted from a volcano and an understanding of the processes of magma evolution in the plumbing system allow evaluating the probable type of eruption in the future.

The main limitation of the petrological and geochemical approaches is that magma samples from active volcanic plumbing systems are generally not available. Although many of the processes of magma evolution can be constrained based on petrological experiments, the major and trace element compositions of rocks and minerals can in many cases not be attributed to any specific process. The contribution of each individual process is often difficult to quantify. Moreover, the insights derived from petrological and geochemical approaches apply to the geological past of the plumbing system only. It is therefore not always straightforward to conclude on the present state.

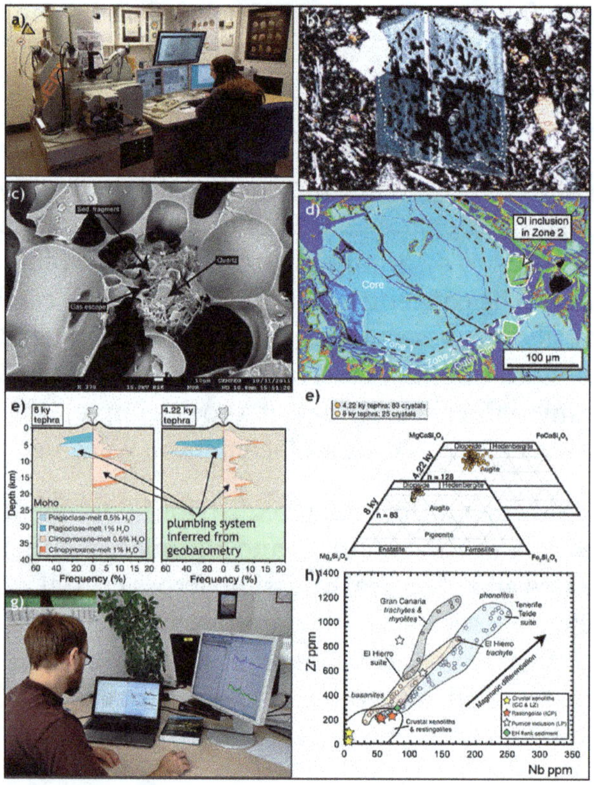

In figure, examples of petrological and geochemical studies of volcanic plumbing systems. (a) High-precision chemical analysis and element mapping using a field emission gun electron probe microanalyser. (b) Thin section of a zoned plagioclase crystal (zonation indicated by dashed line) in basalt of the 2015 Holhraun eruption. Crossed polars. Crystal ca. 1 mm across. Crystal textures can be used to reconstruct processes in the volcanic plumbing system. (c) Scanning electron microscope (SEM) image of vesiculated xenolith erupted offshore El Hierro. (d) False-colour SEM image of a zoned clinopyroxene crystal in lava erupted at Holuhraun, Iceland. Zones and inclusions can be used to reconstruct processes in the volcanic plumbing system. (e) Results of geobarometric modelling of plagioclase and clinopyroxene in tephras from Katla volcano, Iceland. (f) Mineralogy of pyroxene crystals in Katla tephras. (g) Analysing trace-element contents in igneous rocks. (h) Zr versus Nb plot of igneous rocks and xenoliths from the Canary Islands, Spain. Trace element compositions can be used to understand the origin of, and relationships between, rock groups.

Geophysical Studies of Volcanic Plumbing Systems

Geophysical studies of volcanic plumbing systems employ a variety of methods that detect and quantify either the physical properties of different geomaterials, such as magma versus solid rock or igneous versus sedimentary rocks, or the effects of active physical processes, such as seismicity caused by the movement of magma through the crust. Geophysical methods are therefore applied to study both active volcanoes and extinct subsurface or eroded plumbing systems.

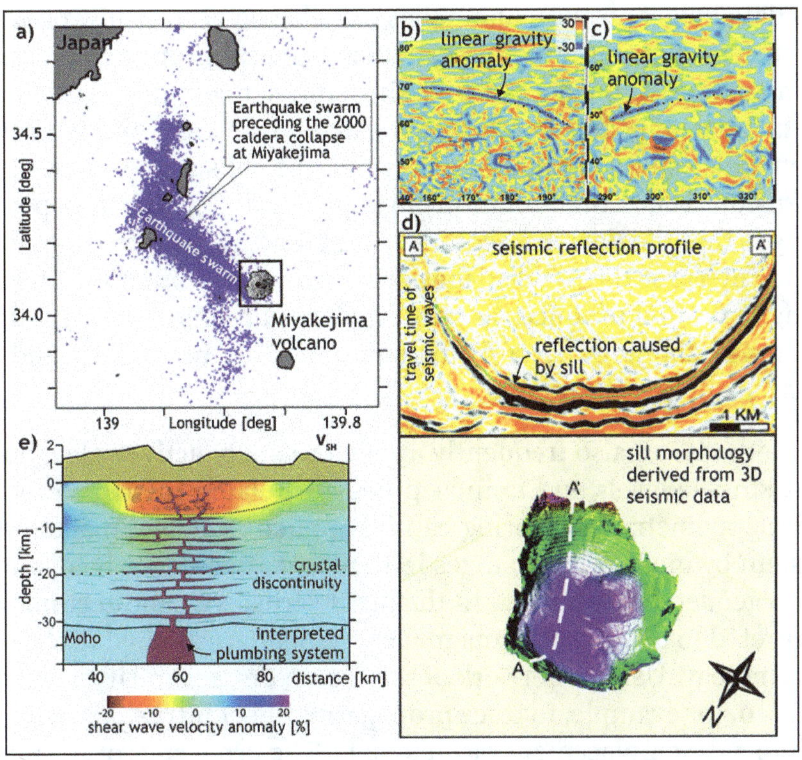

In figure, examples of geophysical studies of volcanic plumbing systems. (a) Seismicity related to the emplacement of a dyke preceding caldera collapse in Miyakejima volcano, Japan, in 2010. (b) and (c) Linear gravity anomalies (marked by black dots) interpreted to be related to dykes on the Moon. (d) Above: Vertical section of a 3D seismic dataset showing the signal related to a sill. Below: 3D seismic image of the same saucer-shaped sill. (e) Volcanic plumbing system of the Toba caldera inferred from seismic tomography.

In order to characterise physical properties of different materials in the crust, a wide range of geophysical methods, such as gravimetry, magnetometry, and electrics, are commonly applied to map volcanic plumbing systems and associated hydrothermal systems. These methods detect anomalies in the Earth's gravimetric, magnetic, and electrical resistivity fields. The anomalies are caused by the presence of rock types with properties contrasting to those of the surrounding country rocks, for example, higher or lower density, resistivity etc. Depending on the type and size of target, for example, mapping a granitic intrusion, and the expected depth range of the target, appropriate methods, and acquisition techniques are chosen. Instruments may be mounted on airplanes and satellites or installed in arrays at the Earth's surface. Acquired geophysical data can then be used to map the physical properties of the Earth's crust. Such maps allow locating the extent and map the outline of magmatic intrusions, which is extremely useful for large-scale reconnaissance and mapping in poorly exposed or inaccessible areas, such as moons or other planets. The spatial resolution of the data mainly depends on the contrast in physical properties between the target and the country rock. The acquired data on geophysical anomalies can also be used in inversion models that infer the subsurface distribution of a particular physical property by reproducing the pattern and magnitude of the anomaly making simple assumptions about the geology, for example, homogeneity and isotropy of the involved rocks. For instance, the results of inversion models can be applied to determine the subsurface shape and volume of a granitic intrusion. The main limitation of geophysical mapping at depth using inversion modelling is that the data are acquired at the Earth's surface (more or less in 2D) is then used to interpret the 3D distribution of physical properties. Moreover, inversion modelling that often produces as very similar solutions in spite of different input parameters. Hence, the number of fitting solutions may be infinite, and thus, model interpretations are often non-unique.

Geophysical methods are also frequently applied to study active volcanoes in order to locate magma storage levels and to infer processes related to magma movement. For instance, microgravimetry monitoring can be applied to detect magma flow into the plumbing system by monitoring changes in the local gravimetric field, a method often combined with geodetic monitoring. In the brittle crust, volcano-tectonic earthquakes that are interpreted to be related to magma movement are studied in the subdiscipline of volcano seismology. Using a network of seismometers, seismicity in active volcanoes can be monitored, for example, to trace propagating sheet intrusions, which gives valuable insight into dyke emplacement mechanisms. The properties of seismic waves (frequencies, wave forms, etc.) can be interpreted to derive information about processes such as the rise of magma in a conduit or the intrusion of a cryptodome. Travel times of seismic waves through the crust beneath a volcano can be used as input parameters for inversion models, a technique called seismic tomography. The seismic velocity structure of a volcano can be used to map subsurface areas with an increased percentage of melt, potentially corresponding to magma reservoirs. However, the detection limit of seismic tomography for magma bodies is on the order of several hundred metres, depending on the seismic array and acquisition conditions on the one hand and the structure of the individual volcano on the other. Therefore, most parts of the volcanic plumbing system, such as average sized dykes and sills, may be undetectable using seismic tomography.

During the last decade, seismic reflection and refraction have been extensively implemented to image volcanic plumbing systems in sedimentary basins. While all geophysical methods described earlier employ passive measurements of natural rock properties, seismic reflection and refraction methods represent active geophysical surveying: acoustic waves are sent into the subsurface and reflected and refracted at interfaces between rocks of contrasting impedance. Thus, the resulting seismic images are a direct echography of subsurface structures and not the result of data inversion. Initially designed for studying the structure of sedimentary basins and hydrocarbon exploration, seismic data proved essential to document the presence of voluminous sill and laccolith complexes in numerous basins worldwide. On seismic images, igneous intrusions are very prominent because of the strong impedance contrast between igneous and sedimentary rocks. The main advantages of seismic data are that they can image the shapes of entire intrusions in three dimension, which can potentially constrain magma flow directions, and it is possible to constrain the intrusion-scale deformation induced by magma emplacement and intrusion-fault interactions. The main limitations of seismic data are the limited spatial resolution (about 20 to 40 m), the limited possibility to image subvertical features, such as dykes, seismic artefacts that produce interpretable features that do not exist, and the availability of the seismic data, as they are often kept confidential by oil companies.

Geophysical methods are thus powerful tools to study physical properties and signals produced by processes of volcanic plumbing systems. Despite their limitations, these methods are often the only way to derive information about the location and properties of plumbing systems at depth.

Geodetic Studies of Volcanic Plumbing Systems

Volcano observatories monitor active volcanoes with a combination of techniques, including seismic networks, temperature, and gas monitoring, as well as geodetic techniques. Geodetic monitoring of the deformation of the Earth's surface relies mainly on ground-based tiltmeters, GPS networks, and satellite-based Interferometry of Synthetic Aperture Radar (InSAR). A network of GPS stations has to be deployed and maintained in the field and typically delivers daily data for each station. Continuous monitoring can provide surface deformation data with an accuracy of about 1 mm/yr, but the data quality strongly depends on the station network density. On the other hand, InSAR records surface deformation in the line of sight of the satellite at 10–40 days intervals. So, InSAR can reach accuracies of less than 1 cm for a stack of interferograms. Apart from disturbances due to vegetation and glaciation, InSAR allows geodetic monitoring of volcanoes worldwide, irrespective of accessibility. Thanks to InSAR monitoring data, we now know of more than 140 currently deforming volcanoes.

The corrected and processed surface deformation data are interpreted using geodetic volcano-deformation modelling, which reproduces the pattern and magnitude of the surface signal with analytical models based on fluid and solid mechanics. The results of geodetic models have been used to interpret volcano deformation in many volcanoes worldwide and have led to a better understanding of magma movements. Even though surface deformation in volcanoes is not always related to an imminent eruption, geodetic monitoring, and modelling have become standard tools for eruption forecasting.

Geodetic models are analytical or numerical solutions that comprise three components: A. a deformation source representing the VPS, B. a static process in the deformation source, and C. the model crust bounded by the free surface and characterised by a rheological law. The combination of A, B, and C produces deformation of the model surface, which is compared to the measured surface deformation. As the solution of geodetic models with highly different input parameters can produce very similar results, the geodetic modeller is left with an infinite number of possible, non-unique solutions. Non-linear inversion is thus applied to select likely source parameters, to estimate uncertainties, and to compare the goodness-of-fit of different models. When comparing the fit of models, the simplest one, which reproduces the recorded data best, is usually preferred, even though it may be geologically implausible for the studied volcano.

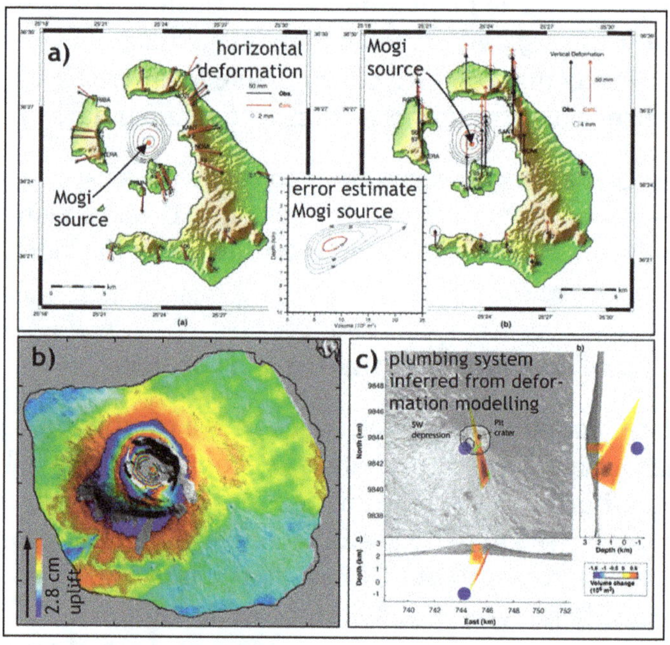

In figure, examples of geodetic studies of volcanic plumbing systems. (a) GPS data (black arrows) and results of Mogi models (red arrows) of ground deformation of Santorini. The Mogi source (red dot) is given with confidence intervals in per cent. (b) Post-eruptive deformation of Fernandina volcano, Galapagos islands, measured with InSAR. (c) Structure of the volcanic plumbing system of Nyamulagira volcano, Congo, inferred from InSAR and stress modelling.

Since it is impossible to quantify the uncertainties of geodetic model results on active geological processes that occur in the subsurface, it is important to constrain the models by using model input parameters provided by depth and location of earthquakes related to moving magma, earthquake "shadows" around magma reservoirs, or by volcanic eruption volumes and volcanic plume heights. However, these methods miss small- and medium-sized magma chambers, as well as the shape of the plumbing system. Furthermore, geobarometry may indicate the depth and duration of long-term magma storage, but cannot resolve levels of short-time storage. More importantly, however, the lack of collaborative work between different disciplines that study volcanoes has so far prevented a more realistic interpretation of surface deformation in volcanoes. So in practice, deformation source geometries are often arbitrarily chosen without geological validation.

Laboratory Modelling

The aim of laboratory modelling is to simulate processes in volcanic plumbing systems at manageable scales, such as laboratory lengths (sand boxes of a few metres length maximum) and time scales (minutes to days). In the nineteenth and early twentieth centuries, laboratory models were phenomenological only and used as source of inspiration and/or proof of concepts to demonstrate the existence of a phenomenon. The main limitations of these models were their applicability to geological systems. Laboratory models entered a new era after Hubbert introduced the scaling theory that established scale relationships between laboratory-scale model and geological-scale system. However, these scale relationships address the similarity between the two systems only, that is, they cannot be used to understand the physics behind the modelled processes. Hence, the development of the dimensional analysis concept within the field of physics was a breakthrough for laboratory modelling. Dimensional analysis identifies fundamental physical laws, so-called scaling laws, that are simple relationships between dimensionless parameters, that is, their validity is scale independent. Combining dimensional analysis and similarity principles to laboratory models allow the modeller to unravel the fundamental physical laws and apply them to geological-scale volcanic plumbing systems.

Successful implementation of laboratory models requires the use of model materials with relevant properties. Classic rock analogues are gels and loose sand to simulate elastic rocks and cohesion-less Coulomb (i.e. frictional) rocks, respectively, which are two end-member behaviours of natural rocks. Recently, new materials of more complex rheology have been successfully used. Cohesive granular materials, some of which have variable cohesion, account for the complex elastoplastic properties of the brittle upper crust, and can simulate both mode I (tensile) and mode II (shear) fracturing. Gels offer the possibility to study the viscoelastic properties of the Earth's crust. On the other hand, many different fluids of diverse rheology and viscosity have been used as magma analogues. Most noticeable is the implementation of two-phase fluids to model the dynamics of magma-bubble suspensions. These materials offer the possibility to model the natural complexity of geological systems, and hence, the physics of the complex processes in volcanic plumbing systems is studied.

Another recent enhancement of laboratory models has been the implementation of various monitoring methods that allow for more quantitative data acquisition and analysis. The most commonly used methods are Particle Image Velocimetry (PIV) and Digital Image Correlation (DIC) and stereo-photogrammetry to monitor deformation of surfaces, as well as X-ray scanners to monitor in situ model interiors. The resulting quantitative data have become essential for (i) constraining the physical laws governing the modelled processes and (ii) integrating laboratory results with quantitative geological, geophysical and geodetic data.

The main limitations of laboratory models to be considered include the scale gap between the laboratory and geological systems, which is often critical and hard to fully constrain. The full characterization of model materials is not straightforward, even though it is essential, and requires a solid fundamental mechanics background. For example, cohesive and cohesion-less granular materials exhibit elastic properties that are difficult to measure, though they are crucial for the dynamics of the brittle crust. Another challenge is connected to recording the modelled processes, which often requires advanced techniques, such as laser or X-ray scanning, which are very costly and heavy to handle. Moreover, stress fields in granular material experiments cannot be measured, although stress distribution is a key factor on the dynamics and evolution of VPS. Finally, first-order assumptions need to be made due to technical limitations of model systems, neglecting for instance thermal effects and chemical evolution of magmas.

Theoretical Modelling

In order to assess first-order scaling laws governing a process in volcanic plumbing systems, theoretical (or analytical) models are employed. Theoretical models solve analytically or semianalytically the mathematical equations governing the studied processes. During the mathematical derivations, the first-order scaling parameters commonly appear spontaneously, such that the first-order effect of the key physical parameters is obvious, and it is possible to identify whether distinct physical behaviours can be expected. Pioneering theoretical models of hydraulic fractures, for example, are the foundation for most subsequent models of dyke or laccolith emplacement.

Although theoretical models are powerful, they require an advanced level of mathematical skill, which is often beyond the training level of most geoscientists. Moreover, theoretical models exhibit limitations, such that the solution of the equations requires a large number of simplifying assumptions. As a consequence, many geoscientists are not aware of the assumptions behind the mathematical models and extrapolate the model results beyond their domain of validity. Furthermore, geoscientists often accept the assumptions underlying the theoretical models as rules, such that the assumptions are rarely questioned, in spite of contradicting evidence from nature. For example, it is commonly assumed that dykes and sills are hydraulic fractures emplaced in purely elastic host rocks, although clear field, geophysical, and laboratory evidence of first-order inelastic deformation accommodating their emplacement are frequently observed.

Numerical Modelling

Numerical modelling uses mathematical equations to simulate a simple process or a combination of several processes. Using numerical codes and software, which have often been developed for materials science applications, the studied system is subdivided into subsystems, so-called elements, and the equations solved for each element. Hence, numerical models can overcome some of the main limitations of theoretical models by accounting for boundaries of complex shapes, complex heterogeneities, etc. The models can take into account material properties, temperatures and pressures and calculate stresses and strains, as well as changes in temperature, pressure, and material properties. Furthermore, numerical models can account for static, quasistatic, and transient processes. Therefore, in a transient model, it is possible to calculate at each time step, the stress field, flow field, and/or temperature field. This represents a major advantage with respect to laboratory models, in which the material properties and processes within the model are challenging to measure and monitor. Another advantage is that the scales of numerical models can be directly set as the scale of geological systems. As computational power is the main constraint on how complex a numerical model may be, it is easy to systematically vary model parameters, such as the material properties, a major advantage compared to laboratory modelling.

There are two main types of numerical models with fundamentally different approaches, continuum models and discrete models. Continuum models solve the fundamental equations of continuum mechanics, such as Stokes and Navier-Stokes equations for fluid flow, Hook's law for elastic deformation, the heat equation calculating heat diffusion and/or advection, and Darcy's law for porous flow. Various solving methods exist, such as finite difference, finite element, and finite volume. In volcanic plumbing system research, continuum models are commonly used to model the elastic deformation induced by magmatic intrusions, magma flow within intrusions, thermal impacts of intrusions on their host rock, thermal convection within magma reservoirs, and recently magma emplacement. The main limitation of continuum models is that they cannot simulate the formation of new discontinuities, such as fractures or dykes, which is why the geometry of discontinuities has to be prescribed.

Discrete models calculate the behaviour of a pack of particles that interact with each other. Some main methods are discrete element models (DEM) and Lattice Boltzman (LBM). The particle interactions are represented by bounds with elastic, friction, or fluid-like properties and strengths. Discrete models are very valuable to simulate processes that create discontinuities, such as fractures, and they appear very suitable to simulate caldera collapse, dyke, sill, and laccolith emplacement. The main limitation of discrete models is to link the particle-scale interactions (e.g. bound stiffness, bound strength) with the bulk properties of the particle packing (Young's modulus, cohesion, internal friction), requiring heavy systematic calibration before running the models and interpreting the results.

Challenge: Integrating the Methods

Since our training and level of experience fundamentally influence how we interpret geological problems, different studies of volcanic plumbing systems that employ methods of different disciplines often lead to contradictory conclusions. For instance, shallow magma storage beneath Katla volcano has been suggested based on seismic and geodetic data, whereas petrological studies so far found evidence for lower- to mid-crustal storage only. Such contrasting results may lead to fundamentally different interpretations when it comes to volcanic risk and hazard assessment and may therefore have unforeseeable consequences. However, these contrasting results may be due to the limitations of individual methods that commonly assess a fraction of the complex magmatic dynamics only, a discrepancy that is hardly possible to assess for decision makers.

The disciplinary boundaries in plumbing system research in part reflect the historical development of the Earth sciences in general. During the twentieth century, each of the disciplines involved in the study of magma transport and storage has become more specialised, and new disciplines, such as volcano geodesy, have emerged. Methods have become more and more sophisticated, revealing the complexity of individual, and the interplay between, physical and chemical processes at scales ranging from the size of the crystal lattice to the thickness of the lithosphere. While specialisation and methodological progress naturally continue, the true challenge of the twenty-first century is to overcome the boundaries of our disciplines in order to truly assess the complexity of volcanic plumbing system dynamics.

Tools to Monitor Volcanoes

Assessing Seismicity

The simplest and cheapest way to monitor a volcano is with seismometers. In an area with several volcanoes that have the potential to erupt (e.g., the Squamish-Pemberton area), a few well-placed seismometers can provide us with an early warning that something is changing beneath one of the volcanoes, and that we need to take a closer look. There are currently enough seismometers in the Lower Mainland and on Vancouver Island to provide this information.

A seismometer installed in 2007 in the vicinity of the Nazco Cone, B.C.

If there is seismic evidence that a volcano is coming to life, more seismometers should be placed in locations within a few tens of kilometers of the source of the activity. This will allow geologists to determine the exact location and depth of the seismic activity so that they can see where the magma is moving.

Detecting Gases

Water vapor quickly turns into clouds of liquid water droplets and is relatively easy to detect just by looking, but CO_2 and SO_2 are not as obvious. It's important to be able to monitor changes in the composition of volcanic gases, and we need instruments to do that. Some can be monitored from a distance (from the ground or even from the air) using infrared devices, but to obtain more accurate data, we need to sample the air and do chemical analysis. This can be achieved with instruments placed on the ground close to the source of the gases, or by collecting samples of the air and analyzing them in a lab.

Measuring Deformation

A GPS unit installed at Hualalai volcano, Hawaii. The dish-shaped antenna on the right is the GPS receiver. The antenna on the left is for communication with a base station.

There are two main ways to measure ground deformation at a volcano. One is known as a tiltmeter, which is a sensitive three-directional level that can sense small changes in the tilt of the ground at a specific location. Another is through the use of GPS (global positioning system) technology. GPS is more effective than a tiltmeter because it provides information on how far the ground has actually moved — east-west, north-south, and up-down.

By combining information from these types of sources, along with careful observations made on the ground and from the air, and a thorough knowledge of how volcanoes work, geologists can get a good idea of the potential for a volcano to erupt in the near future (months to weeks, but not days). They can then make recommendations to authorities about the need for evacuations and restricting transportation corridors. Our ability to predict volcanic eruptions has increased dramatically in recent decades because of advances in our understanding of how volcanoes behave and in monitoring technology. Providing that careful work is done, there is no longer a large risk of surprise eruptions, and providing that public warnings are issued and heeded, it is less and less likely that thousands will die from sector collapse, pyroclastic flows, ash falls, or lahars. Indirect hazards are still very real, however, and we can expect the next eruption like the one at Laki in 1783 to take an even greater toll than it did then, especially since there are now roughly eight times as many people on Earth.

Volcano Gravimetry

Gravimetry is a discipline overlapping the fields of geophysics, geodesy, and geodynamics that deals with the geometry, properties, and changes of the earth gravity field by means of its observation, analysis, and interpretation.

When considering volcanic hazards, the knowledge of the following elements is essential: the geological past of a volcano, the geological structure of a volcano and its tectonic setting, and the subsurface processes taking place inside a restless or awakening volcano or deep underneath it within the crust and upper mantle. None of these three elements can be observed directly. The geological past of a volcano is inferred from its products left from the past eruptions and from the morphology of the volcano and its surroundings. The chemical and mechanical composition of the products testifies about the violence, size, and character (such as explosive versus effusive) of past eruptions. Geochemical, petrological, and morphological methods help recover such information. The geological structure of a volcano is not accessible to the naked eye either. Geophysical methods interpreting observations made on the surface, which are sensitive to the distribution of physical properties (parameters) of the volcano edifice, and the crust or uppermost mantle underneath, are applied to compile this knowledge. Seismic methods, gravimetry, magnetics, electrical soundings, induction, self- potential, and ground penetrating radar are examples of such exploration. Similar is the situation with acquiring knowledge on the subsurface volcanic processes linked with migration of magma and associated volatiles. From data monitored or observed on the surface, we aim at reconstructing the reality happening within the volcano. Subsurface physical and chemical (compositional) changes in the magma reservoirs (or changes in the volcano plumbing system in general), in hydrothermal systems, and within their surrounding rock environment, as well as underground mass movement, produce observable geophysical signals on the surface. In the sequel, we shall focus on two such observables: the temporal gravity changes and the vertical displacements (elevation changes) as the vertical component of the deformations of the earth surface.

Surface Deformations and Gravity Changes

Processes associated with the physical and chemical changes in the magma storage system and intruding fresh magma both lead to changes in subsurface stress field producing strain that manifests itself on the surface in terms of surface deformations. The subsurface strain field and changes in the temperature field generate changes in the subsurface density distribution, which is sensed on the surface as microgravity changes. Moreover, the movement (migration) of masses such as magma within preexisting feeding or storage systems, as well as formation of new parts of the plumbing system in terms of dikes, sills, inclined or conical sheets, and rings, produces another component of changes in the subsurface density distribution. The same holds true for the movement of volcanic fluids (or volatiles in general) within the porous rock medium or along cracks and fractures or for any changes in hydrothermal systems as such. Any changes in the subsurface density distribution manifest themselves on the surface as changes in the gravitational attraction vector. Typically, the vertical component is observed on the surface, referred to as temporal gravity changes. Naturally, both the stress field changes and the subsurface density distribution changes work in a concert, implying a simultaneous occurrence of both the surface displacements and the gravity changes. Therefore, it would be natural and advantageous to interpret surface deformations and gravity changes jointly. We must also mention that these two observables are coupled. This is due to the fact that any surface deformations lead to gravity changes caused by the vertical displacement of the topographic surface in the ambient gravity field regardless of the subsurface mass redistribution.

Volcanological experience has proven that the inflation (or deformation in general) of the volcano edifice is an essential precursor of volcanic eruptions. Therefore, the monitoring of surface deformations plays an irreplaceable role in volcanic threat assessment. Surface deformations can be monitored using continuous or repeat survey terrestrial geodetic techniques as well as extraterrestrial techniques, such as differential interferometric synthetic aperture radar (DInSAR) or its permanent scatterer modification (PSInSAR). Gravity changes turned out to be also a valuable tool for predicting the reactivation of a volcano or for studying volcanic unrest. Cases have been reported when gravity changes were detected before seismic or other precursors at a reawakening volcano (e.g.,). Gravity changes can be monitored continuously using tidal relative gravimeters or absolute gravimeters or in a repeat campaign (survey) mode at a network of gravity points (benchmarks, stations) covering a representative area over the volcano.

While the repeat campaign (time-lapse) observations can better capture the spatial extent of the gravity changes, the continuous measurements at a station (or several stations) can pick up the fast changes to which the repeat survey mode is blind in the intra-campaign period. The best, of course, appears to be the combination of the two modes.

Surface deformations are sensitive particularly to source pressure or volume changes. Gravity changes help characterizing the source process by sensing mass changes along with volume changes, thus pointing at the source density changes. Monitoring and interpreting temporal changes in 3D gravity data has been recently referred to as "4D (micro)gravity" or 4D gravimetry. A comprehensive overview of the purpose and benefits of the 4D gravity, both the discrete (time-lapse, repeat survey) and the continuous, illustrated on several cases at specific volcanoes. A special case happens when gravity changes are observed without any significant surface deformations or vice versa. By "significant" we mean within the accuracy threshold of the measuring technique. This strengthens the argument for concurrent monitoring of both the surface deformations and the gravity changes in volcanic areas.

Surface deformations and gravity changes can be interpreted either as stand-alone quantities or jointly. We have already indicated the pitfalls of monitoring and interpreting just one of them and the advantages of monitoring and interpreting them jointly. The (sole or joint) interpretation is based on either modeling or inversion. Modeling is based on assuming a hypothesis about a source and its parameters, simplifying the structural rock environment, and solving the forward problem yielding the model surface deformations and model gravity changes. By iteratively changing the source parameters or the hypothesized model source, modeling is carried out until a reasonable fit ("match") is achieved between observed and modeled quantities (displacements and gravity changes). When a fit is reached, the model source with its parameters represents one of the possible solutions representing the volcanic process taking place at depth. Two approaches to modeling are at hand: analytical and numerical.

In analytical modeling, significant simplifications are applied to the source as such (its geometry) and to the structural environment: neglecting the topography and subsurface geologic structure (assuming the environment to be a homogenous isotropic elastic or viscoelastic half-space). Presuming simple geometries of the sources (such as magma chambers, dikes, and sills) leads to closed-form analytical expressions linking source parameters and the observable surficial quantities. Hence, analytical modeling can give quick results and first impressions about the nature and characteristics of the source process. On the contrary, the oversimplifications used in analytical modeling can lead to severe distortions of the obtained results with the risk of misleading the interpreter.

In numerical modeling, the simplifications of the analytical approach are remedied. The numerical approach takes into account the effect of topography, structural discontinuities and inhomogeneity, and even rheology. Such approach calls for applying finite element or boundary element methods (FEM, BEM).

Inversion works differently: a nonlinear inverse problem is solved by means of computing the source parameters directly from the observed surficial data. This gravimetric inverse problem is nonunique and ill-posed. The methods to solve it are the same as those in potential field inverse problems.

Some volcanic areas, especially calderas, display during their restlessness a special behavior—they produce systematic trends in their gravity/height ratios. By analogy, these ratios are typically referred to as gravity/height gradients. When the plots of gravity changes versus elevation changes show systematic trends clearly distinguished by two linear boundaries [free air gradients (FAGs) and Bouguer gradients], they are considered as signatures of the dominating typified subsurface processes: magma rejuvenation or drainage, vesiculation or degassing, and water table rise or fall.

The mathematical source solutions (model solutions) obtained from either modeling or direct inversion must be associated with processes taking place in the volcano edifice or even deeper in the crust or lithospheric mantle and furthermore often with a volcanic threat or a prediction about an eruption. There is an everlasting problem present with the monitoring and interpretation of geodetic (deformations) and gravimetric (gravity changes) data in volcanic areas: they are neither easy nor cheap to observe—their temporal as well as spatial resolution is never high enough to satisfy the inversion and interpretation demands. This will become particularly evident when we shall further below discuss the novel approaches to gravity inversion that require the input data given on an equidistant (regular) and dense enough grid. When the spatiotemporal gravity changes do not have a sufficient resolution and accuracy, it becomes hard, nay impossible, to discriminate among the possible sources of volcanic unrest and possibly draw reasonable conclusions about the threat of impending eruption.

Decomposition of Superimposed Gravity Signals

Several natural (physical) phenomena produce a change in gravity observed at a point (benchmark) on the earth surface: changes in atmospheric attraction (due to mainly atmospheric pressure changes), tidal effects such as solid earth tides and ocean (loading) tides, hydrological effects such as groundwater table level changes, changes in snow or ice cover, soil moisture, volcanic effects associated with changes in magma storage (state of magma and magma transport), as well as magmatic volatiles/fluids. All of these signals (components)—gravitational effects of the individual contributions—are superimposed to form the gross observed gravity changes. When we aim at studying a particular geodynamic phenomenon, such as magma replenishment into an existing plumbing system of a volcano, a magma intrusion, or eventually magma vesiculation/degassing in a shallow chamber or vent, using gravimetry, we must be able to decompose the composite gravity signal. In other words, we need to remove (correct for) all unwanted signals. The net signal, stripped of any unwanted components, is typically referred to as "residual gravity changes".

First, we remove the contribution of the environment: gravity change components imposed by the atmosphere and tides. Then, we need to deal with the hydrological component due to pluvial water (precipitation). This turns out to be a cumbersome correction to handle in practice, the size of which often is at the level of the net signal chased after. Magma is associated with its volatiles and brines. Hydrothermal systems often reside above the magma storage systems. Physical changes (such as temperature and pressure) in both the magmatic and the hydrothermal systems have gravimetric signatures that may be difficult, nay impossible, to discern. Similar situation may arise with the gravimetric signatures associated with the transport (mobility) of magma and hydrothermal fluids.

This has become obvious in monitoring and interpreting unrest at calderas. Sometimes, researchers come to diverse conclusions interpreting the same data (gravity changes and surface deformations). Some interpretations favor a magma intrusion process (whereas the opposing ones prefer a hydrothermal reasoning. Naturally, these two source processes may act simultaneously complementing each other, being referred to as hybrid unrest.

For the purpose of inverting and interpreting the net signal, the residual gravity changes can be compiled using the following decomposition:

$$\Delta g^{res} = \Delta g^{obs} - \Delta g^{ext} - \Delta g^{inst} - \Delta g^{w} - \Delta g^{def} - \Delta g^{surf},$$

where the superscripts denote the following components of the gravity change: residual (res), observed (obs), external (ext), instrumental (inst), hydrological (w), deformation induced (def), and surficial (surf).

By external component, we mean environmental effects composed of tidal and atmospheric effects. Tidal effects comprise solid earth (body) tides and ocean loading effects. Atmospheric effects are gravity changes induced by pressure and temperature changes in the atmosphere (atmospheric attraction and loading effects). These effects are known and respective corrections have been published.

As instrument/survey effects, we consider the drift of gravimeters and adjustment of redundant measurements. We want to highlight one particular effect that must not be neglected in monitoring and interpreting time-lapse gravity change observations. Various gravimeters have various heights of the sensor, measured from the bottom of the instrument. In addition, various plates or tripods can be used, implying various heights of the bottom of the instrument above a benchmark (gravity station, gravity point). If various types of gravimeters or plates/tripods are used within an epoch survey or between epochs, observed gravity must be reduced to the elevation of the benchmark. Otherwise, significant systematic errors could be introduced to the residual gravity changes. To reduce the meter reading to the ground, actual vertical gradient of gravity (VGG) is needed.

Under hydrological effects, we mean the changes in the groundwater table level as well as in the soil/rock moisture (due to precipitation or drought). Strictly speaking to correct for this environmental signal, detailed 3D modeling needs to be performed, which is typically unachievable due to lack of required input data and/or knowledge about the near surface (or subsurface) geological structure (porosity, impermeable interfaces, etc.). Various approximations and estimates are used in practice; data on water table from wells are used wherever available. The simplest approximation is a planar Bouguer effect of a water table level change. The hydrological correction is not the subject of our interest here, so we refer the reader to published works. Deformation-induced topographic effects (DITEs) are of our high concern here.

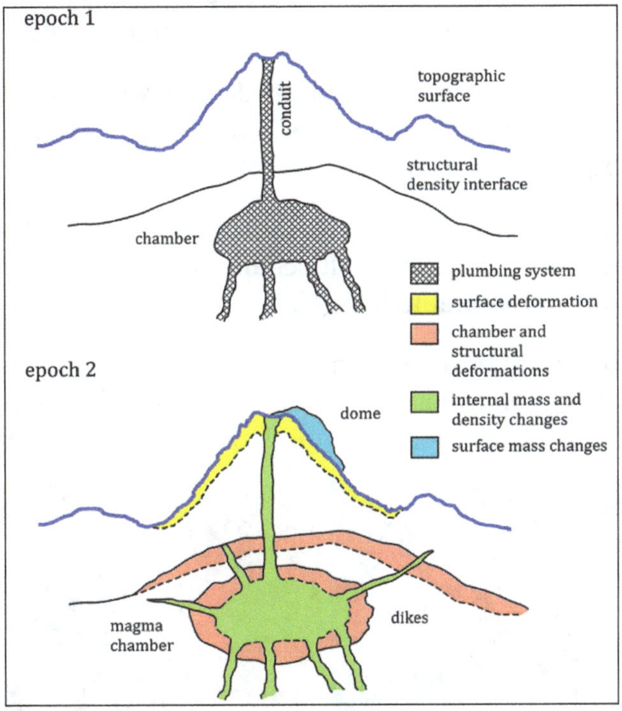

Schematic sketch of the sources contributing to the composite gravity changes.

Under the surface effects, we mean gravitational effects of mass changes taking place on the surface, but unrelated to surface deformation, such as magma extrusion, dome growth, dome collapse, lava flow, lahar, and flank collapse. These masses can be surveyed on the surface, their shapes and sizes can be digitized, and their gravitational effects can be computed by a numerical realization of the Newton volume integral for the vertical component of the attraction and subsequently subtracted. This may be achieved with the help of photogrammetry, laser scanning, LIDAR, etc. These corrections are out of scope of our interest here.

The residual gravity changes are consequently related to the subsurface volcanic processes associated with magma migration (rejuvenation or drainage of existing reservoirs and feeders, intrusions, and propagation along new paths, such as diking), hydrothermal fluid (volatile) migration, density changes due to physical and chemical (compositional) changes of the residing magma (cooling, heating, vesiculation, degassing, fractional crystallization, differentiation, mixing, partial melting, etc.), and density changes of the geological structure of the edifice due to stress-induced strain field (including the deformation of density interfaces) and due to temperature field changes via thermal expansivity. All of these changes represent the net sought signal, which in its nature is a fairly complex composite signal itself. This component of the gravity changes, the "residual gravity changes", is the subject of inversion or forward modeling and of subsequent interpretation of magma-related processes.

We want to take a look at some of the recent developments and contributions to the field of volcano gravimetry in the sequel. We will turn our attention to three topics:

- Revisiting the DITEs on gravity changes dealing with the coupling between vertical displacements (elevation changes) and gravity changes, attempting to propose a more rigorous and accurate way of handling them.
- Taking a look at the benefits arising from applying a novel inversion methodology, as applied to time-lapse gravity changes, to which we shall briefly refer as the "Prutkin inversion methodology".
- Probing the possibilities of applying a novice inversion methodology, developed by Pohánka, based on n-harmonic (polyharmonic) functions, referred to as the "harmonic inversion method", to time-lapse gravity changes.

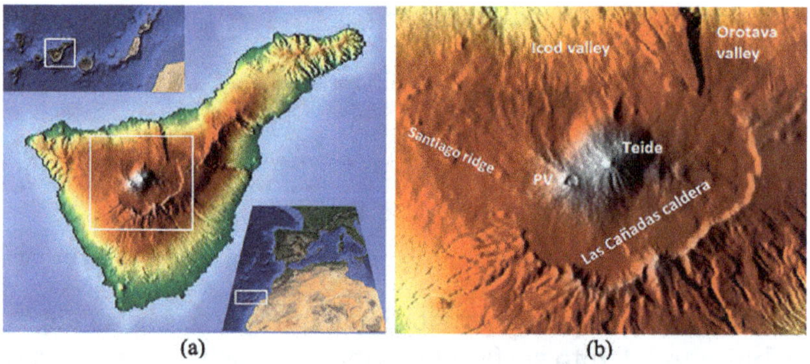

(a) Location of the Canary Islands and Tenerife and
(b) the caldera and the twin stratovolcanoes Teide and Pico Viejo (PV).

For all the three above topics, we chose the Central Volcanic Complex (CVC) of Tenerife, Canary Islands, as our case study playground, on which we ran either synthetic simulations based on high-resolution digital elevation models and Mogi point sources or processed and inverted the gravity changes observed during the 2004 to 2005 volcanic unrest. The selected area has a suitable (significant and jaggy) topography for our simulations, as it comprises a caldera at the average elevation of roughly 2000 m asl, within which twin stratovolcanoes Teide and Pico Viejo are located reaching altitudes 3718 and 3135 m asl, respectively. Among other unrest indicators, spatiotemporal gravity changes were observed at the CVC at 14 benchmarks of a rapid reaction network between May 2004 and July 2005. No statistically significant areal surface deformation (either inflation or deflation) was observed accompanying these gravity changes. The observed gravity changes were corrected for tidal and hydrological effects. These point gravity data were taken as input data in both our inversion approaches. They were interpolated (extrapolated) onto a regular (equidistant) grid.

Coupling between Vertical Deformations and Gravity Changes

Let us consider the gravitational effects imposed by the deformation of the topographic surface, i.e., by vertical displacements (elevation changes), illustrated, for instance, on inflation (uplift).

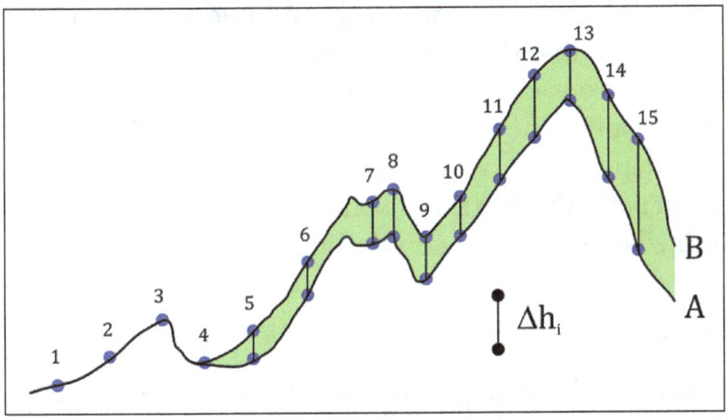

Schematic sketch of the DITEs on the gravity change. Benchmarks 5 to 15, unlike 1 to 4, experienced a significant uplift (Δhi). "A" denotes the predeformation topographic surface, whereas "B" denotes the postdeformation surface.

During the deformation (uplift), the gravity station (benchmark) is vertically displaced together with the topographic surface. We note that the subsurface density (geological) structure (including discontinuity interfaces) is deformed, too, but we treat this effect as part of the residual gravity changes to be inverted and interpreted. The reason for this is that, unlike the topographic surface, which is known (described by a DEM), the geological structure is usually unknown or poorly known. We can hypothetically split the ground deformation effect into two subsequent stages (steps) with their respective effect components. In the first step, we move (lift up) the benchmark vertically from its original position (P) on the predeformation topographic surface to its new postdeformation position (P*) in the ambient gravity field (in "free air") without deforming (inflating) the topographic surface (topographic masses) yet. In the second step, we move (deform, inflate) the topographic masses. Thus, we can write for these two separated components:

$$\Delta g^{def}\left(P^*\right) = \Delta g^{\mathrm{FAE}}\left(P^*\right) + \Delta g^{\mathrm{TDE}}\left(P^*\right),$$

where Δg^{def} is the DITE; Δg^{FAE} is the gravity change due to the vertical displacement of the benchmark in ambient gravity field (in "free air"), in the sequel referred to as the "free air effect" (FAE); and Δg^{TDE} is the gravitational effect (vertical component of the attraction vector) of the topographic masses that were displaced into the volumetric domain between the predeformation and the postdeformation topographic surfaces. We have coined this effect the "topographic deformation effect" (TDE). Already from the sketch of figure, we can intuitively deduce that, at benchmarks 1 to 4, which experienced no uplift (no significant elevation changes), the FAE will be zero, whereas the TDE will not be zero. The masses displaced into the topographic deformation rind will impose gravitational effect not only on benchmarks 5 to 15, which have undergone nonzero elevation changes, but also on benchmarks 1 to 4, which experienced no uplift, zero elevation changes. This is an important insight. The size of the TDE at a specific benchmark will depend on its vertical position relative to, and horizontal distance from, the deformation shell (rind) and on the volume and shape of the deformation shell. Below, we shall examine the TDE by numerical simulations. As a consequence, the DITE might be nonzero even at benchmarks with zero elevation changes. This may hint on the possible insufficiency of expressions of the DITE that linearly depend on the vertical displacement, i.e., those being a product of some sort of a gradient (such as Bouguer or free air) and the vertical displacement. These issues associated with the DITE will be examined by numerical simulations further below.

FAE and its Approximations

The FAE amounts to the vertical displacement of the benchmark($\Delta h(P)$)times the true (in situ) VGG:

$$\Delta g^{\text{FAE}}(P^*) = (\partial g / \partial h)^o (P) * \Delta h(P),$$

Inevitably, the VGG must be observed in situ at the benchmark (hence the superscript "o"). This can be practically achieved by relative gravimeters observing in a so-called tower mode, i.e., on the ground and at a certain height above the benchmark, such as 1 m, on a tripod.

If the VGG is not observed in situ, and its value remains unknown, it is usually (one could say habitually) approximated by the "theoretical FAG", also called the "normal FAG", which is the constant average vertical gradient of normal gravity at the surface of the normal reference ellipsoid. In the sequel, we shall abbreviate the normal FAG as NFAG (NFAG=-308.6 μGal/m). Observations indicate that the approximation of the true VGG by the NFAG can introduce a relative error of up to 88% in rugged terrain of alpine mountains, such as the High Tatras of Slovakia, or 77% at the CVC of Tenerife as shown.

In figure TNFAG values at the CVC of Tenerife. Minima (in absolute value) correspond with sharp concave topofeatures (valleys, gullies, drains), whereas maxima (in absolute sense) with sharp convex topofeatures (peaks, ridges, pillars, crests, and ribs). Max = −70, min = −481, mean = −315, SD = 34 (μGal/m).

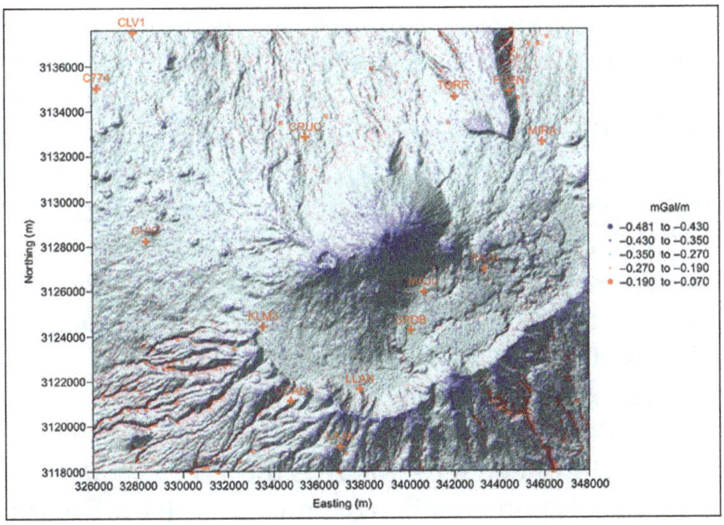

Therefore, we propose a better approximation of the true VGG, in case it is not observed, than the NFAG. In rugged terrain regions, the strongest (primary) contribution to the VGG comes from the terrain (topography and eventually also bathymetry for near shore points) from within the near vicinity of the benchmark. This contribution can be modeled (adopting a constant topodensity) using precise high-resolution digital terrain models. The NFAG can be refined (corrected for) using this topographic contribution to the VGG. In the sequel, we shall abbreviate the "topocorrected NFAG" as "TNFAG". We show numerical values of the TNFAG for the CVC of Tenerife in figure. The computational method and the used DEM are described in. The topographic contribution (correction) amounts in the CVC area of Tenerife to 77% (in relative sense) of the NFAG itself, which is highly significant. This is the amount by which the true VGG may differ from the NFAG in a volcanic area of rugged topography.

The secondary contribution to the true VGG comes from the underground geological structure (density anomalies) of the earth. This contribution typically remains unmodeled, whereas the structure remains unknown. Alternatively, Rymer and Berrino et al. propose an approximation (of the VGG), a refinement to NFAG, based on the Bouguer anomaly map of the area. The contribution of geology to FAG in volcanic areas with significant (rugged) topography is expected to be smaller than that of topography, although, in flat areas (such as some calderas), the situation may be vice versa.

TDE and its Approximations

The TDE is the gravitational effect (attraction) of the masses between the topographic surfaces before and after the deformation (within the topographic deformation shell) on the observed gravity change at each benchmark of the survey. It must be computed by numerical evaluation of the Newton integral (for vertical component of the attraction vector) over this volumetric domain with assumed mean constant topographic density (ρoρo representing the near surface rock environment).

The TDE is in volcanogravimetric applications commonly approximated by a Bouguer plate effect referred to as planar "Bouguer deformation effect" (BDE): $\Delta g^{TDE}(P^*) \approx \Delta g^{BDE}(P^*)$, where $\Delta g^{BDE}(P^*) = 2\pi G_{\rho 0*} \Delta h(P) G$ being the gravitational constant. Alternatively, the TDE is approximated by a spherical Bouguer term, which is a combination of three dilation effects respective to a Mogi point source model $\Delta g^{TDE}(P^*) \approx (4/3)\pi G_{\rho 0*} \Delta h(P)$. For a detailed discussion about these approximations, as well as how the Bouguer gradients respective to these two Bouguer approximations relate to the gravity/height gradients, we refer the reader to. In both (planar and spherical) Bouguer approximations of the TDE, the BDE depends linearly on the displacement of the benchmark. This fact alone is a clear indication that the BDE might not be a sufficient approximation of the TDE: for benchmarks with zero vertical displacement, yet close enough to the deformed topographic masses to sense their attraction (at the order of 10 μGal), the BDE will be zero, whereas the TDE will not. This difference between the TDE and its BDE approximation (combined with the departure of the in situ VGG from the NFAG) explains the occurrence of observed data at many volcanoes that in terms of gravity/height plots often dramatically and erratically deviate from the linear trends of the FAG or Bouguer-corrected FAG (BCFAG).

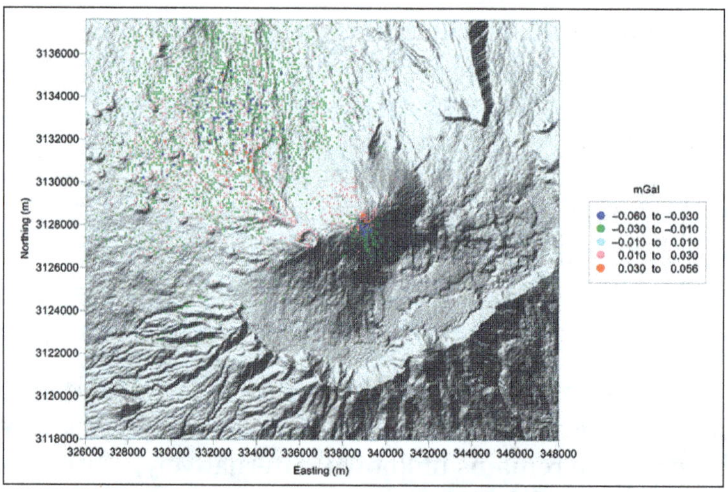

Difference between the TDE and its approximation by the planar BDE, both respective to a synthetic displacement field simulated at the CVC of Tenerife. Max = 56, min = −60, mean = −1, SD = 6 (μGal).

To illustrate the size and spatial characteristics of the TDE—most of all that it significantly departs from its Bouguer approximation (BDE), we computed the TDE at the CVC of Tenerife for a synthetic deformation (surface displacement) field generated by two (shallow and deep) Mogi sources and compared it to the planar BDE for the same displacement field by displaying the difference between the two in Figure. The shallow Mogi source was located at the depth of 500 m roughly below the Teide summit, scaled to have a displacement magnitude of 1 m, whereas the deep source was located at the depth of 6 km roughly 5 km to the northwest of the twin stratovolcanoes, scaled to have a displacement magnitude of 50 cm. For such significant surface deformations, the difference between the TDE and the BDE is also significant. In future simulations, we plan to compute the TDE also for smaller deformation fields to assess its deviation from the BDE approximation for smaller displacements.

Prutkin Inversion Methodology

The methodology under a brief working title "Prutkin inversion methodology" refers to a modular inversion methodology that consists of several modules or, in other words, several processing and inversion steps or algorithms. This modularity leads to producing several diverse inversion solutions that all equally well satisfy (mathematically) the input data (time-lapse gravity changes in the case of volcano gravimetry). The methodology was originally developed and applied to structural geophysical studies at global, regional, and even local scales. We have applied it recently to temporal gravity changes and the results seem promising.

This methodology is used to be referred to as the "method of local corrections", which is not telling, as the name refers to the last algorithm of the methodology only. That is why we refer to it here loosely as the "Prutkin methodology". The methodology consists of the following blocks or algorithms: (a) trend removal (optional), where trend is defined as a 2D harmonic function and computed from the input data; (b) decomposition into shallow/deep (or more depth interval) field components (optional) using a triple harmonic continuation numerical procedure; (c) 3D line segments approximation of the sources, where the gravitational effect of these sources approximates the input gravity data; (d) field decomposition based on the individual or grouped line segment effects; and (e) solutions in terms of potato-shaped (closed compact star-convex) source bodies and/or contact surfaces (density discontinuity interfaces). All of these data processing and inversion steps are described mathematically. When the methodology is applied to (time-lapse) gravity changes instead of gravity anomalies, the represent subsurface time-lapse (temporal) density changes instead of density anomalies. For instance, the potato-shaped bodies are volumetric domains of a homogenous density change.

A pilot study was carried out on gravity changes observed at Mayon Volcano, Philippines. This study revealed that the methodology is applicable to time-lapse gravity changes observed during unrest or volcanic activity. The second study was performed on gravity changes observed during the 2004/5 unrest on Tenerife. This study manifested the true potential of the methodology. It produced a more revealing gravimetric picture of the unrest compared to the previous gravimetric interpretations. The decomposition of the observed gravity changes into shallow and deep fields lead to the unveiling of the deeper source that was interpreted as magmatic. This was possible only thanks to the unique feature of the methodology—the depth-wise decomposition of the signal based on the triple harmonic continuation. Previous interpretations were not capable of detecting this deeper source characterized as of magmatic origin. The sources pertinent to the shallow field were interpreted as hydrothermal fluids. Consequently, the 2004/5 unrest on Tenerife was evaluated, based on the gravimetric interpretation, as hybrid (due to the presence of both magmatic and hydrothermal sources). Seismic constraints were adopted to assist and strengthen the gravimetric interpretation.

The great advantages of the Prutkin inversion methodology dwell in its modularity and versatility. Several combinations can be put together in processing the input data. Several combinations in decomposing the input signal can be introduced. This leans on whether trend removal is performed or not, whether the input signal is decomposed into shallow and deep fields (or even several depth components) or not, and what depth level is chosen for its decomposition. If the input data allow the approximation of sources by several line segments, additional decomposition of the input signal (field) is possible by separating the signal respective to the individual line segments or various groups of line segments. In addition, each field component can be inverted either in terms of a density contrast contact surface (interface) or in terms of 3D star convex compact homogenous ("potato-shaped") source body. Moreover, a priori density contrasts (differential densities), as well as a priori depths for the asymptotic planes of the interfaces, have to be selected, as they are free parameters. The choice of various density contrasts of various source bodies and/or interfaces creates additional variability in the sizes and shapes of the inversion solution sources. Consequently, the versatility and modularity of the methodology results in a possibly rich set of diverse inversion solutions. All of these solutions will equally well satisfy the input data (in terms of the forward problem). However, not all of them will be equally meaningful or reasonable in terms of geological, tectonic, geodynamic, or volcanological considerations. The great advantage of the presented methodology dwells in its ability to produce plentiful diverse solutions to the gravimetric inverse problem that can be at a later stage discriminated based on independent geophysical constraints (such as seismic) or cognition stemming from other earth science disciplines (such as geochemistry and petrology) or alternatively stemming from the experience of the interpreter. An illustration of such variability of gravimetrically acceptable diverse solutions can be found for the Kolarovo gravity anomaly in the Danube Basin presented in. The same applies to gravimetric inversion of time-lapse gravity changes.

Harmonic Inversion Method

The harmonic inversion method was developed by Pohánka. It has been applied to gravity data in structural geophysical studies. We were curious about its applicability to temporal (time-lapse) gravity changes, so we stepped to testing it on the gravity changes observed during the 2004/5 unrest on Tenerife. We used the same data set as in the Prutkin inversion methodology to be able to compare inversion results of both methodologies.

The gravimetric inverse problem is nonunique. It has infinitely many solutions. Some special solutions can be sought, characterized by being smooth. This methodology searches for subsurface 3D solutions satisfying three conditions: The solution is a n-harmonic function, where n is a small integer (n=3, n=4); The solution has an extremum preservation property for a point source field, i.e., the main extremum of the solution coincides with the location of the point source generating the input gravitational field;

and the solution can be expressed as a linear integral transformation of the input gravity field. The tetraharmonic (n=4) solutions are called "characteristic density" or shortly "chi density". Such harmonic solutions are smooth and unrealistic, as they do not reflect subsurface geological settings. Therefore, in the subsequent step, these harmonic solutions are transformed to piece-wise homogenous density distributions by an automated iterative procedure, in which the harmonic solution (as a 3D subsurface function computed from the 2D surficial input data) serves as a mediator controlling the iterations in terms of the goodness of fit. The piece-wise homogenous solution consists of a set of homogenous closed-form source bodies of specific density contrasts with respect to an ambient constant density. Each source body can have its own density contrast.

The current version of the methodology works with a triharmonic function (n=3) called "quasi-gravitation", which serves the needs of the mediator in the iterative process resulting in the piece-wise constant subsurface density distribution, given as a set of homogenous source bodies. Density contrasts (differential densities) of these source bodies have to be selected a priori by the interpreter before running the iterative procedure. The iterative procedure, which modifies the shapes and sizes of the individual source bodies, runs until a satisfactory (preselected sufficient) fit is obtained between the quasi-gravitation of the found source bodies and the quasi-gravitation of the observed input gravity data. This fit guarantees also the fit between the surface gravity of the found source bodies and the observed input gravity. Regarding all the details about this inversion methodology based on n-harmonic functions, we refer the reader to. When inverting gravity, the differential densities of the solution represent structural density anomalies. When inverting temporal (time-lapse) gravity changes, the differential densities represent time-lapse density changes.

References

- Volcanology, science: britannica.com, Retrieved 5 June, 2019
- Wright, Laura. "Geotimes - November 2000: New Notes". Www.geotimes.org. Archived from the original on 8 August 2017. Retrieved 23 March 2018
- Reading-volcanic-arcs, chapter, geo: courses.lumenlearning.com, Retrieved 6 July, 2019
- Walker, George P. L. (1999-12-01). "Volcanic rift zones and their intrusion swarms". Journal of Volcanology and Geothermal Research. 94 (1–4): 21–34. Doi:10.1016/S0377-0273(99)00096-7
- Studying-volcanic-plumbing-systems-multidisciplinary-approaches-to-a-multifaceted-problem, updates-in-volcanology-from-volcano-modelling-to-volcano-geology, books: intechopen.com, Retrieved 7 August, 2019
- Reading-tools-to-monitor-volcanoes, chapter, geo: courses.lumenlearning.com, Retrieved 8 August, 2019
- Rheological properties of basaltic lavas at sub-liquidus temperatures: laboratory and field measurements on lavas from Mount Etna". Cat.inist.fr. Archived from the original on November 2, 2015. Retrieved October 27, 2016
- Recent-developments-and-trends-in-volcano-gravimetry, updates-in-volcanology-from-volcano-modelling-to-volcano-geology, books: intechopen.com, Retrieved 9 January, 2019

4
Volcanic Eruptions

A volcanic eruption is said to occur when hot materials like lava, molten rocks, gaseous compounds, dust, etc. are ejected from the Earth's crust. Some of the different types of volcanic eruptions are Hawaiian eruption, Strombolian eruption, Submarine eruption, Plinian eruption, etc. This chapter closely examines these types of volcanic eruptions as well as the causes of volcanic eruption.

Volcanic Eruption is the sudden occurrence of a violent discharge of steam and volcanic material. Deep within the Earth it is so hot that some rocks slowly melt and become a thick flowing substance called magma. Since it is lighter than the solid rock around it, magma rises and collects in magma chambers. Eventually, some of the magma pushes through vents and fissures to the Earth's surface. Magma that has erupted is called lava.

Some volcanic eruptions are explosive and others are not. The explosivity of an eruption depends on the composition of the magma. If magma is thin and runny, gases can escape easily from it. When this type of magma erupts, it flows out of the volcano. A good example is the eruptions at Hawaii's volcanoes. Lava flows rarely kill people because they move slowly enough for people to get out of their way. If magma is thick and sticky, gases cannot escape easily. Pressure builds up until the gases escape violently and explode. A good example is the eruption of Washington's Mount St. Helens. In this type of eruption, the magma blasts into the air and breaks apart into pieces called tephra. Tephra can range in size from tiny particles of ash to house-size boulders.

Explosive volcanic eruptions can be dangerous and deadly. They can blast out clouds of hot tephra from the side or top of a volcano. These fiery clouds race down mountainsides destroying almost everything in their path. Ash erupted into the sky falls back to Earth like powdery snow. If thick enough, blankets of ash can suffocate plants, animals, and humans. When hot volcanic materials mix with water from streams or melted snow and ice, mudflows form. Mudflows have buried entire communities located near erupting volcanoes.

Causes of Volcanic Eruptions

Magma forms deep in the Earth's crust and in the upper parts of the mantle. In these areas, the temperature and pressure are very high. Changes in pressure and temperature can cause magma to form.

Part of the upper mantle is made of very hot, solid rock. The rock is so hot that it can flow, like soft chewing gum, even though it is solid. If rock of this temperature were found at the Earth's surface, it would be molten, or melted. The rock in the mantle does not melt because it is under high pressure. This pressure is produced by the weight of the rock above the mantle.

In the figure below, the curved line shows the melting point of a rock. The melting point is the temperature at which the rock melts for a certain pressure.

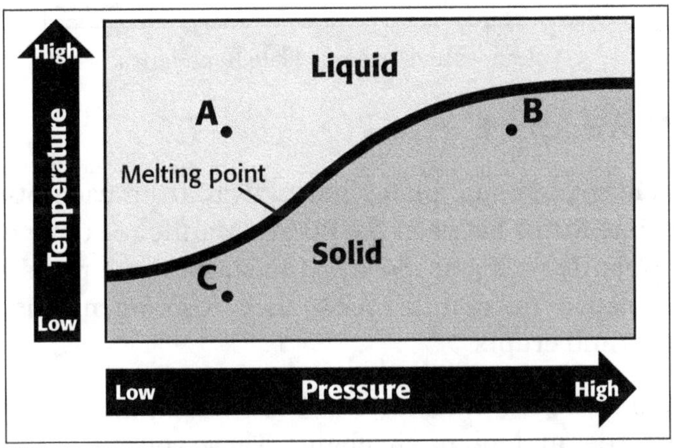

The curved line shows the melting point of the rock. Rock with the temperature and pressure of point A is liquid. Rock at the same temperature but higher pressure (B) is solid. Rock at the same pressure but lower temperature (C) is also solid.

Magma Formation in the Mantle

The temperature of the mantle is fairly constant. Magma usually forms because of a decrease in pressure. Therefore, a lot of magma forms at the boundary between separating tectonic plates, where pressure decreases. Magma is less dense than the solid rock it forms from. Therefore, it rises toward the surface and erupts.

Where do Volcanoes Form?

The locations of volcanoes give clues about how volcanoes form. The figure below shows the locations of some of the world's major active volcanoes. The map also shows the boundaries between tectonic plates. Most volcanoes are found at tectonic plate boundaries. For example, there are many volcanoes on the plate boundaries surrounding the Pacific Ocean. Therefore, the area is sometimes called the Ring of Fire.

Remember that tectonic plate boundaries are areas where plates collide, separate, or slide past one another. Most volcanoes are found where plates move together or apart. About 15% of active volcanoes on land form where plates separate, and about 80% form where plates collide. The remaining few volcanoes on land are found far from tectonic plate boundaries.

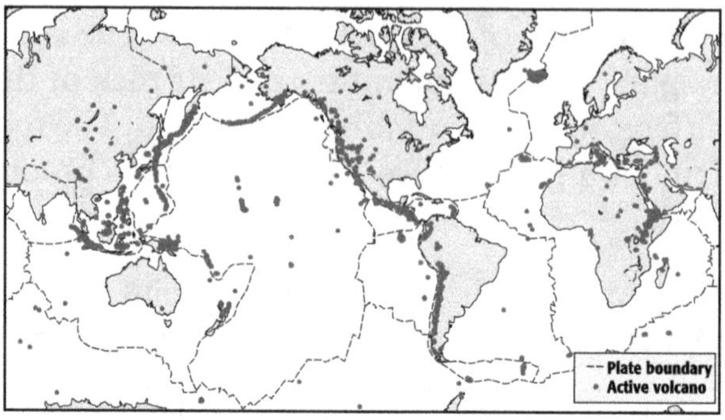

Volcanoes and Tectonic Plate Boundaries.

Where Plates Move Apart

At a divergent boundary, tectonic plates move away from each other. A set of deep cracks called a rift zone forms between the plates. Mantle rock moves upward to fill in the gap. When the mantle rock gets close to the surface, the pressure decreases. The decrease in pressure causes the mantle rock to melt, forming magma. The magma rises through the rift zones and erupts.

Most divergent boundaries are on the ocean floor. Lava that flows from undersea rift zones produces volcanoes and mountain chains. These volcanoes and mountain chains are called mid-ocean ridges. The mid-ocean ridges circle the ocean floor.

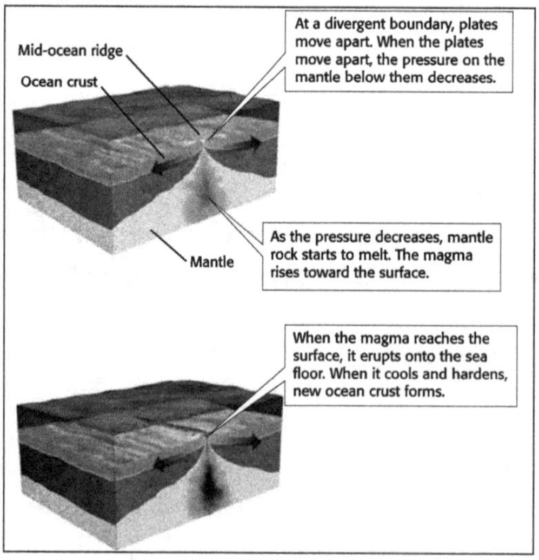

Where Plates Move Together

At a convergent boundary, the tectonic plates collide. When an oceanic plate collides with a continental plate, the oceanic plate slides under the continental plate. This is called subduction. The oceanic crust sinks into the mantle because it is more dense than the continental crust.

As the ocean crust sinks, the temperature and pressure on it increase. Because the ocean crust forms below the ocean, the rock contains a lot of water. The heat and pressure on the ocean crust cause this water to be released.

The water mixes with the mantle rock above the oceanic plate. When the mantle rock mixes with water, it can melt at a lower temperature. The mantle rock begins to melt at the subduction zone. The magma rises to the surface and erupts as a volcano.

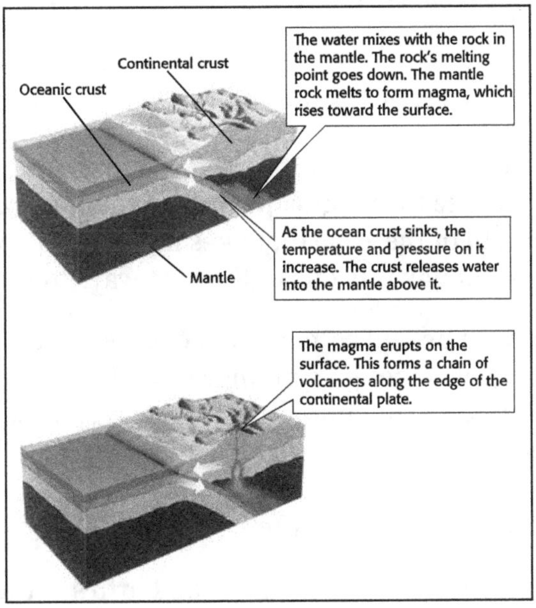

In the Middle of Plates

Although most volcanoes form at plate boundaries, not all volcanoes form there. Some volcanoes, such as the Hawaiian Islands, form at hot spots. Hot spots are places on the Earth's surface where volcanoes form far from plate boundaries. Most scientists think that hot spots form above hot columns of mantle rock called mantle plumes. Some scientists think that hot spots form where magma rises through cracks in the Earth's crust.

Long chains of volcanoes are common at hot spots. One theory to explain this is that a mantle plume stays in one place while the plate moves over it. Another theory states that hot-spot volcanoes occur in long chains because they form along cracks in the Earth's crust. Scientists are not sure which of these theories is correct. It is possible that some hot spots form over plumes, but others form over cracks.

How can we Predict Volcanic Eruptions?

Scientists cannot always predict when and where a volcano will erupt. However, by studying ancient and modern volcanoes, scientists have been able to identify some signs that an eruption may happen.

One feature that scientists use to predict whether an eruption will happen is the state of the volcano. Geologists put volcanoes into three groups based on how active they are:

- Extinct volcanoes have not erupted in recorded history and probably will not erupt again.
- Dormant volcanoes are currently not erupting, but they may erupt again.
- Active volcanoes are currently erupting or show signs of erupting in the near future.

Small Quakes and Volcanic Gases

Most active volcanoes produce small earthquakes as the magma within them moves upward. This happens because the magma pushes on the rocks as it rises. In many cases, the number and strength of these earthquakes increases before a volcanic eruption. Therefore, monitoring earthquakes is one of the best ways to predict an eruption.

Scientists also study the volume and composition of gases given off by the volcano. Just before an eruption, many volcanoes give off more gas. The composition of the gas may also change before an eruption. By monitoring the gases, scientists can predict when an eruption may happen.

Measuring Slope and Temperature

As magma rises before an eruption, it can cause the Earth's surface to swell. The side of a volcano may even bulge. Scientists can use an instrument called a tiltmeter to measure the slope of the volcano's sides. Changes in the slope can indicate that an eruption is likely.

One of the newest methods for predicting volcanic eruptions involves using satellite images. Satellites can record the surface temperatures at and around volcanoes. As magma rises, the surface temperature of the volcano may increase. Therefore, an increase in surface temperature can indicate that an eruption is likely.

Hawaiian Eruption

A Hawaiian eruption is a type of volcanic eruption where lava flows from the vent in a relatively gentle, low level eruption; it is so named because it is characteristic of Hawaiian volcanoes. Typically they are effusive eruptions, with basaltic magmas of low viscosity, low content of gases, and high temperature at the vent. Very small amounts of volcanic ash are produced. This type of eruption occurs most often at hotspot volcanoes such as Kīlauea on Hawaii's big island and in Iceland, though it can occur near subduction zones (e.g. Medicine Lake Volcano in California, United States) and rift zones. Another example of Hawaiian eruptions occurred on the island of Surtsey in Iceland from 1964 to 1967, when molten lava flowed from the crater to the sea.

Hawaiian eruptions may occur along fissure vents, such as during the eruption of Mauna Loa in 1950, or at a central vent, such as during the 1959 eruption in Kīlauea Iki Crater, which created a lava fountain 580 meters (1,900 ft) high and formed a 38-meter cone named Puʻu Puaʻi. In fissure-type eruptions, lava spurts from a fissure on the volcano's rift zone and feeds lava streams that flow downslope. In central-vent eruptions, a fountain of lava can spurt to a height of 300 meters or more (heights of 1600 meters were reported for the 1986 eruption of Mount Mihara on Izu Ōshima, Japan).

Hawaiian eruptions usually start by the formation of a crack in the ground from which a curtain of incandescent magma or several closely spaced magma fountains appear. The lava can overflow the fissure and form ʻaʻā or pāhoehoe style of flows. When such an eruption from a central cone is protracted, it can form lightly sloped shield volcanoes, for example Mauna Loa or Skjaldbreiður in Iceland.

Petrology of Hawaiian Basalts

The key factors in generating a Hawaiian eruption are basaltic magma and a low percentage of dissolved water (less than one percent). The lower the water content, the more peaceful is the resulting flow. Almost all lava that comes from Hawaiian volcanoes is basalt in composition. Hawaiian basalts that make up almost all of the islands are tholeiite. These rocks are similar but not identical to those that are produced at ocean ridges. Basalt relatively richer in sodium and potassium (more alkaline) has erupted at the undersea volcano of Lōʻihi at the extreme southeastern end of the volcanic chain, and these rocks may be typical of early stages in the "evolution" of all Hawaiian islands. In the late stages of eruption of individual volcanoes, more alkaline basalt also was erupted, and in the very late stages after a period of erosion, rocks of unusual composition such as nephelinite were produced in very small amounts. These variations in magma composition have been investigated in great detail, in part to try to understand how mantle plumes may work.

Safety

Hawaiian eruptions are considered less dangerous than other types of volcanic eruptions, due to the lack of ash and the generally slow movement of lava flows. However, they can still cause injuries or deaths.

In 1993, a photographer attempting to take pictures of a lava ocean entry died, and several tourists were injured, when a lava bench collapsed. In 2000, two people were found dead near a lava ocean entry from Kīlauea, likely killed by laze. Sulfur Dioxide emissions can also be fatal, especially to people suffering from respiratory ailments. In 2018 lava spatter from Kīlauea broke a man's leg, and another 23 people on a tour boat were injured by a steam explosion at a lava ocean entry from the same eruption.

Strombolian Eruption

A Strombolian eruption is a type of volcanic eruption with relatively mild blasts, having a volcanic explosivity index of about 1 to 3. Strombolian eruptions consist of ejection of incandescent cinders, lapilli, and lava bombs, to altitudes of tens to a few hundreds of metres. The eruptions are small to medium in volume, with sporadic violence. This type of eruption is named for the Italian volcano Stromboli.

The Italian vulcanologist Giuseppe Mercalli studied eruptions at Stromboli and Vulcano in 1888–1890, and observed that the characteristic features of eruptions were different between the two. To distinguish between them, Mercalli defined Strombolian eruptions as “Mildly explosive at discrete but fairly regular intervals of seconds to minutes”.

The tephra typically glows red when leaving the vent, but its surface cools and assumes a dark to black colour and may significantly solidify before impact. The tephra accumulates in the vicinity of the vent, forming a cinder cone. Cinder is the most common product; the amount of volcanic ash is typically rather minor.

The lava flows are more viscous, and therefore shorter and thicker, than the corresponding Hawaiian eruptions; it may or may not be accompanied by production of pyroclastic rock.

Instead the gas coalesces into bubbles, called gas slugs, that grow large enough to rise through the magma column, bursting near the top due to the decrease in pressure and throwing magma into the air. Each episode thus releases volcanic gases, sometimes as frequently as a few minutes apart. Gas slugs can form as deep as 3 kilometers, making them difficult to predict.

Strombolian eruptive activity can be very long-lasting because the conduit system is not strongly affected by the eruptive activity, so that the eruptive system can repeatedly reset itself.

Monogenetic cones usually erupt in the Strombolian style. For example, the Parícutin volcano erupted continuously between 1943–1952, Mount Erebus, Antarctica has produced Strombolian eruptions for at least many decades, and Stromboli itself has been producing Strombolian eruptions for over two thousand years. The Romans referred to Stromboli as the “Lighthouse of the Mediterranean”.

Vulcanian Eruption

A Vulcanian eruption is a type of volcanic eruption characterized by a dense cloud of ash-laden gas exploding from the crater and rising high above the peak. They usually commence with phreatomagmatic eruptions which can be extremely noisy due the rising magma heating water in the ground. This is usually followed by the explosive clearing of the vent and the eruption column is dirty grey to black as old weathered rocks are blasted out of the vent. As the vent clears, further ash clouds become grey-white and creamy in colour, with convolutions of the ash similar to those of Plinian eruptions.

The term Vulcanian was first used by Giuseppe Mercalli, witnessing the 1888–1890 eruptions on the island of Vulcano. His description of the eruption style is now used all over the world. Mercalli described Vulcanian eruptions as "Explosions like cannon fire at irregular intervals." Their explosive nature is due to increased silica content of the magma. Almost all types of magma can be involved, but magma with about 55% or more silica (e.g. basaltic andesite) is most common. Increasing silica levels increase the viscosity of the magma which means increased explosiveness.

Characteristics of Vulcanian Eruption

Vulcanian eruptions display several common characteristics. The mass of rock ejected during the eruption is usually between 10^2 - 10^6 tonnes and contains a high proportion of non-juvenile material (> 50%). During active periods of volcanic activity, intervals between explosions vary from less than 1 minute (e.g. Anak Krakatoa) to about a day. Pyroclastic flows are also common features of this type of eruption. The gas streaming phase of Vulcanian eruptions are characterised by discrete cannon-like explosions. These expulsions of gas can reach supersonic velocities resulting in shock waves.

The Gran Cratere, Vulcano. A sense of scale is provided by the tourist visible near the centre of the crater.

The tephra is dispersed over a wider area than that from Strombolian eruptions. The pyroclastic rock and the base surge deposits form an ash volcanic cone, while the ash covers a large surrounding area. The eruption ends with a flow of viscous lava. Vulcanian eruptions may throw large metre-size blocks several hundred metres, occasionally up to several kilometres.

Vulcanian eruptions are dangerous to persons within several hundred metres of the vent. Volcanic bombs are common products of this type of eruption. These are initially molten blobs of lava, which rapidly cool into blocks often 2 to 3 m across. At Galeras, a Vulcanian eruption ejected bombs which struck several volcanologists who were in the crater, some of whom died or suffered severe injuries.

Example of Vulcanian Eruption

The 11 September 1930 eruption of Stromboli was a Vulcanian eruption. It started at 08:10 hours (local), when ash was vented for about 10 minutes. Then at 09:52 two incredibly powerful explosions occurred which shook the whole island. Blocks were hurled about 2 km. These fell out of the sky smashing through buildings etc. A tsunami 2 to 2.5 m high was generated. By 10:40 the explosive phase of the eruption was over. Expulsion of lava followed, this flowed down the Sciara del Fuoco, lasting into the night. At the same time incandescent scoria flowed down the Vallonazzo Valley and entered the sea near Piscità.

It is believed that water entered due to a partial collapse of the conduit. The water flashed into steam and took the easiest "escape route" via the open conduit. Expansion by contact with the molten magma generated the two very large explosions.

There were six deaths. Four fishermen died at sea when the avalanches of hot scoria caused the sea to become very disturbed. One person was killed in Stromboli village by falling blocks, and the sixth was killed by the tsunami.

Subglacial Eruption

Subglacial eruptions, those of ice-covered volcanoes, result in the interaction of magma with ice and snow, leading to meltwater formation, jökulhlaups, and lahars. Flooding associated with meltwater is a significant hazard in some volcanic areas, including Iceland, Alaska, and parts of the Andes. Jökulhlaups, glacial outburst floods, have been identified as the most frequently occurring volcanic hazard in Iceland, with major events where peak discharges can reach 10, 000 – 100, 000 m^3/s occurring when there are large eruptions beneath glaciers.

It is important to explore volcano-ice interactions to improve our ability to effectively monitor these events and to undertake hazard assessments. This is particularly relevant given that subglacial eruptions have recently demonstrated their ability to cause widespread impact, with the ash cloud associated with Iceland's Eyjafjallajökull eruption resulting in significant impacts to aviation across Europe.

Deception Island Eruption, Antarctica

Given that subglacial eruptions occur in often sparsely populated regions, they are not commonly observed or monitored; thus timings and sequences of events for an eruption of this type are poorly constrained. Research of the 1969 Deception Island eruption demonstrates that the impact of a subglacial eruption is not limited purely by glacier thickness, but that the pre-volcanic ice structure and densification (proportion of impermeable ice) play a role as well. In this case, even though the glacier was thin, a large jökulhlaup was observed as the glacier was largely made up of impermeable (unfractured) ice with a sudden supraglacial flood once the cavity has reached capacity. The resulting flood severely damaged buildings on the island, with complete destruction of a British scientific station.

Gjalp Eruption, Iceland

Over a period of 13 days, 3 km^2 of ice was melted with erupted magma fracturing into glass to form a 7 km long and 300 m high hyaloclastite ridge under 750 m of ice. Meltwater flowed along a narrow basal glacier bed into a subglacial lake for five weeks, before being released as a sudden flood, or jökulhlaup. Although it has been proposed that subglacial volcanism may play a role in the dynamics of West Antarctic ice streams by supplying water to their base, for Iceland's Gjalp eruption, no rapid basal sliding was observed at the regional scale, with the formation of ice cauldrons over eruptive fissures due to the sudden removal of mass at the base.

Research demonstrated that for warm-based glaciers, the effects of subglacial volcanic eruptions are localised, with eruptions forming deep depressions and causing jökulhlaups. For there to be significant changes in the extent and shape of an ice sheet, extensive subglacial volcanism would be required, melting a considerable fraction of the total ice volume over a short period of time.

Eyjafjallajökull Eruption, Iceland

In the first two days of the eruption, ice cauldrons were formed over the volcanic vents. Radar images reveal the development of these cauldrons in a 200 m thick ice cover within the summit caldera. They can also be used to document the subglacial and supraglacial passage of meltwater away from the eruption site. Research shows the eruption breached the ice surface four hours after the initial eruption onset, whilst meltwater release was characterised by accumulation and subsequent drainage, with most of the volcanic material in the ice cauldrons being drained in hyperconcentrated floods.

Phreatomagmatic Eruption

Phreatomagmatic eruptions are volcanic eruptions resulting from interaction between magma and water. They differ from exclusively magmatic eruptions and phreatic eruptions. Unlike phreatic eruptions, the products of phreatomagmatic eruptions contain juvenile (magmatic) clasts. It is common for a large explosive eruption to have magmatic and phreatomagmatic components.

Mechanisms

Several competing theories exist as to the exact mechanism of ash formation. The most common is the theory of explosive thermal contraction of particles under rapid cooling from contact with water. In many cases the water is supplied by the sea, for example with Surtsey. In other cases the water may be present in a lake or caldera-lake, for example Santorini, where the phreatomagmatic component of the Minoan eruption was a result of both a lake and later the sea. There have also been examples of interaction between magma and water in an aquifer. Many of the cinder cones on Tenerife are believed to be phreatomagmatic because of these circumstances.

The other competing theory is based on fuel-coolant reactions, which have been modeled for the nuclear industry. Under this theory the fuel (in this case, the magma) fragments upon contact with a coolant (the sea, a lake or aquifer). The propagating stress waves and thermal contraction widen cracks and increase the interaction surface area, leading to explosively rapid cooling rates. The two mechanisms proposed are very similar and the reality is most likely a combination of both.

Deposits

Phreatomagmatic ash is formed by the same mechanisms across a wide range of compositions, basic and acidic. Blocky and equant clasts with low vesicle content are formed. The deposits of phreatomagmatic explosive eruptions are also believed to be better sorted and finer grained than the deposits of magmatic eruption. This is a result of the much higher fragmentation of phreatomagmatic eruptions.

Hyaloclastite

Hyaloclastite is glass found with pillow basalts that were produced by non-explosive quenching and fracturing of basaltic glass. These are still classed as phreatomagmatic eruptions, as they produce juvenile clasts from the interaction of water and magma. They can be formed at water depths of >500 m, where hydrostatic pressure is high enough to inhibit vesiculation in basaltic magma.

Hyalotuff

Hyalotuff is a type of rock formed by the explosive fragmentation of glass during phreatomagmatic eruptions at shallow water depths (or within aquifers). Hyalotuffs have a layered nature that is believed to be a result of dampened oscillation in discharge rate, with a period of several minutes. The deposits are much finer grained than the deposits of magmatic eruptions, due to the much higher fragmentation of the type of eruption.

The deposits appear better sorted than magmatic deposits in the field because of their fine nature, but grain size analysis reveals that the deposits are much more poorly sorted than their magmatic counterparts. A clast known as an accretionary lapilli is distinctive to phreatomagmatic deposits, and is a major factor for identification in the field. Accretionary lapilli form as a result of the cohesive properties of wet ash, causing the particles to bind. They have a circular structure when specimens are viewed in hand and under the microscope.

A further control on the morphology and characteristics of a deposit is the water to magma ratio. It is believed that the products of phreatomagmatic eruptions are fine grained and poorly sorted where the magma/water ratio is high, but when there is a lower magma/water ratio the deposits may be coarser and better sorted.

Surface features

Crest of old tuff ring, including part of the maar crater of a monogenetic volcano, Tenerife, Canary Islands. The maar crater has been used for agriculture.

There are two types of vent landforms from the explosive interaction of magma and ground or surface water; tuff cones and tuff rings. Both of the landforms are associated with monogenetic volcanoes and polygenetic volcanoes. In the case of polygenetic volcanoes they are often interbedded with lavas, ignimbrites and ash- and lapilli-fall deposits. It is expected that tuff rings and tuff cones might be present on the surface of Mars.

Tuff Rings

Koko Crater is an old extinct tuff cone in the Hawaiian Island of Oahu.

Tuff rings have a low profile apron of tephra surrounding a wide crater (called a maar crater) that is generally lower than the surrounding topography. The tephra is often unaltered and thinly bedded, and is generally considered to be an ignimbrite, or the product of a pyroclastic density current. They are built around a volcanic vent located in a lake, coastal zone, marsh or an area of abundant groundwater.

Tuff Cones

Tuff cones are steep sloped and cone shaped. They have wide craters and are formed of highly altered, thickly bedded tephra. They are considered to be a taller variant of a tuff ring, formed by less powerful eruptions. Tuff cones are usually small in height. Koko Crater is 1,208 feet.

Examples of Phreatomagmatic Eruptions

Minoan Eruption of Santorini

Fort Rock, an eroded tuff ring in Oregon, USA.

Santorini is part of the Southern Aegean volcanic arc, 140 km north of Crete. The Minoan eruption of Santorini, was the latest eruption and occurred in the first half of the 17th century BC. The eruption was of predominantly rhyodacite composition. The Minoan eruption had four phases. Phase 1 was a white to pink pumice fallout with dispersal axis trending ESE. The deposit has a maximum thickness of 6 m and ash flow layers are interbedded at the top. Phase 2 has ash and lapilli beds that are cross stratified with mega-ripples and dune like structures. The deposit thicknesses vary from 10 cm to 12 m. Phases 3 and 4 are pyroclastic density current deposits. Phases 1 and 3 were phreatomagmatic.

Eruption of Mount Pinatubo

Mount Pinatubo is on the Central Luzon landmass between the South China Sea and the Philippine Sea. The 1991 eruption of Pinatubo was andesite and dacite in the pre-climactic phase but only dacite in the climactic phase. The climactic phase had a volume of 3.7–5.3 km^3. The eruption consisted of sequentially increasing ash emissions, dome growth, 4 vertical eruptions with continued dome growth, 13 pyroclastic flows and a climactic vertical eruption with associated pyroclastic flows. The pre-climactic phase was phreatomagmatic.

Fort Rock, as seen from the ground.

Lake Taupo

The Hatepe eruption in 232+/-12 AD was the latest major eruption at Lake Taupo in New Zealand's Taupo Volcanic Zone. There was minor initial phreatomagmatic activity followed by the dry venting of 6 km^3 of rhyolite forming the Hatepe Plinian Pumice. The vent was then infiltrated by large amounts of water causing the phreatomagmatic eruption that deposited the 2.5 km^3 Hatepe Ash. The water eventually stopped the eruption though large amounts of water were still erupted from the vent. The eruption resumed with phreatomagmatic activity that deposited the Rotongaio Ash.

Effusive Eruption

An 'a'ā lava flow from Mauna Loa during its 1984 eruption.

An effusive eruption is a type of volcanic eruption in which lava steadily flows out of a volcano onto the ground. There are two major groupings of eruptions: effusive and explosive. Effusive eruption differs from explosive eruption, wherein magma is violently fragmented and rapidly expelled from a volcano. Effusive eruptions are most common in basaltic magmas, but they also occur in intermediate and felsic magmas. These eruptions form lava flows and lava domes, each of which vary in shape, length, and width. Deep in the crust, gasses are dissolved into the magma because of high pressures, but upon ascent and eruption, pressure drops rapidly, and these gasses begin to exsolve out of the melt. A volcanic eruption is effusive when the erupting magma is volatile poor (water, carbon dioxide, sulfur dioxide, hydrogen chloride, and hydrogen fluoride), which suppresses fragmentation, creating an oozing magma which spills out of the volcanic vent and out into the surrounding area. The shape of effusive lava flows is governed by the type of lava (i.e. composition), rate and duration of eruption, and topography of the surrounding landscape.

For an effusive eruption to occur, magma must be permeable enough to allow the expulsion of gas bubbles contained within it. If the magma is not above a certain permeability threshold, it cannot degas and will erupt explosively. Additionally, at a certain threshold, fragmentation within the magma can cause an explosive eruption. This threshold is governed by the Reynolds Number, a dimensionless number in fluid dynamics that is directly proportional to fluid velocity. Eruptions will be effusive if the magma has a low ascent velocity. At higher magma ascent rates, the fragmentation within the magma passes a threshold and results in explosive eruptions. Silicic magma also exhibits this transition between effusive and explosive eruptions, but the fragmentation mechanism differs. The 1912 Novarupta eruption and the 2003 Stromboli eruption both exhibited a transition between explosive and effusive eruption patterns.

Basaltic Eruptions

Basaltic composition magmas are the most common effusive eruptions because they are not water saturated and have low viscosity. Most people know them from the classic pictures of rivers of lava in Hawaii. Eruptions of basaltic magma often transition between effusive and explosive eruption patterns. The behavior of these eruptions is largely dependent on the permeability of the magma and the magma ascent rate. During eruption, dissolved gasses exsolve and begin to rise out of the magma as gas bubbles. If the magma is rising slowly enough, these bubbles will have time to rise and escape, leaving a less buoyant magma behind that fluidly flows out. Effusive basalt lava flows cool to either of two forms, ʻaʻā or pāhoehoe. This type of lava flow builds shield volcanoes, which are numerous in Hawaii, and is how the island was and currently is being formed.

Silicic Eruptions

Silicic magmas most commonly erupt explosively, but they can erupt effusively. These magmas are water saturated, and many orders of magnitude more viscous than basaltic magmas, making degassing and effusion more complicated. Degassing prior to eruption, through fractures in the country rock surrounding the magma chamber, plays an important role. Gas bubbles can begin to escape through the tiny spaces and relieve pressure, visible on the surface as vents of dense gas. The ascent speed of the magma is the most important factor controlling which type of eruption it will be. For silicic magmas to erupt effusively, the ascent rate must be $10^{-5}$to 10^{-2}m/s, with permeable conduit walls, so that gas has time to exsolve and dissipate into the surrounding rock. If the flow rate is too fast, even if the conduit is permeable, it will act as though it is impermeable and will result in an explosive eruption. Silicic magmas typically form blocky lava flows or steep-sided mounds, called lava domes, because their high viscosity does not allow it to flow like that of basaltic magmas. When felsic domes form, they are emplaced within and on top of the conduit. If a dome forms and crystallizes enough early in an eruption, it acts as a plug on the system, denying the main mechanism of degassing. If this happens, it is common that the eruption will change from effusive to explosive, due to pressure build up below the lava dome.

Alaskan volcano Novarupta with an effused lava dome at the summit.

Submarine Eruption

Submarine eruptions are those volcano eruptions which take place beneath the surface of water. These occur at constructive margins, subduction zones and within tectonic plates due to hotspots. This eruption style is far more prevalent than subaerial activity. For example, it is believed that 70 to 80% of the Earth's magma output takes place at mid-ocean ridges.

Detection

Submarine eruptions are less studied than subaerial volcanoes due to their inaccessibility. Developments in technology mean that submarine volcanoes can now be studied in greater detail. Despite this progress, understanding is still limited. Mid ocean ridges for example are the most active volcanic systems on Earth but roughly only 5% of their length has been studied in detail.

Initial knowledge of these eruptions came from volcanic rocks being recovered from the ocean floor when repairs were made to the Transatlantic telegraph cable in the

1800s. More recently a variety of techniques have been used to study these eruptions with significant developments being made since 1990. These include the use of remote controlled submersibles which can conduct surveys of the ocean floor. The use of hydrophone networks allows volcanic eruptions to be detected. Submersibles can be sent out in response to this to record the result of the eruption. Other tools have included seismic signals, acoustic waves and high resolution UAV multibeam mapping.

Increasingly, eruptions at greater depths can be observed. For example, an explosive eruption at West Mata in Lau Basin at a depth of 1200 m was studied using submersibles.

Controls on Eruptive Style

There is much variation in the style of submarine eruptions. This changes with a number of variables including magma viscosity, water depth, effusion rate and volatile content. Many studies highlight the effects of pressure which increases with depth. It is believed that increased pressure restricts the release of volatile gases, resulting in effusive eruptions. This is not to say that explosive eruptions do not occur at depth, just that a higher volatile content is required. It has been estimated that at 500 m explosive activity associated with basalts is suppressed, while depths greater than 2300 m would be sufficient to prevent the majority of explosive activity from rhyolite lava.

Shallow Water Eruptions

At shallow depths it is common for submarine eruptions to be explosive due to the reaction between volatiles in the magma and water which generates a significant quantity of steam. These eruptions described as Surtseyan are characterised by large quantities of steam and gas and creating large amounts of pumice. This activity has occurred in many locations. An example is Fukuto-Okanoba near Japan. This activity has been observed for almost a century and causes discoloured water, jets of steam and ash, and pumice is found floating in the surrounding water.

Shallow eruptions can lead to the creation of islands. The most well known is Surtsey in Iceland (1963-1967). Similar island building activity occurs frequently but these are often short lived.

Volatile content is also significant. Magma being transported into the ocean through tunnels may see gases being exsolved before reaching the water and so the eruption is effusive. This has been seen in Hawaii.

Deep Water Eruption

With increased depth there is greater pressure and it is believed that this results in effusive eruptions. There is a variety of evidence, however, which suggests that explosive, pyroclastic activity can occur at depth. This includes observations of Pele's hair and evidence of caldera collapse. This activity is thought to be common at subduction zones due to recycling of the lithosphere. It is not exclusive to these plate margins, occurring at hotspots and ocean ridges. An example is Loihi near Hawaii where both effusive and explosive activity occurs at 2000 m depth.

Two formations associated with submarine eruptions are seamounts and pillow lavas. Pillow lavas are created due to rapid cooling of lava which forms a skin. As more magma is forced into this the skin expands creating a lobe. When this fractures then lava seeps through the gap exposing hot lava to the water and again a skin forms over this: this process is then repeated.

Explosive Eruption

In volcanology, an explosive eruption is a volcanic eruption of the most violent type. A notable example is the 1980 eruption of Mount St. Helens. Such eruptions result when sufficient gas has dissolved under pressure within a viscous magma such that expelled lava violently froths into volcanic ash when pressure is suddenly lowered at the vent. Sometimes a lava plug will block the conduit to the summit, and when this occurs, eruptions are more violent. Explosive eruptions can send rocks, dust, gas and pyroclastic material up to 20 km (12 mi) into the atmosphere at a rate of up to 100,000 tonnes per second, traveling at several hundred meters per second. This cloud may then collapse, creating a fast-moving pyroclastic flow of hot volcanic matter.

Stages of an Explosive Eruption

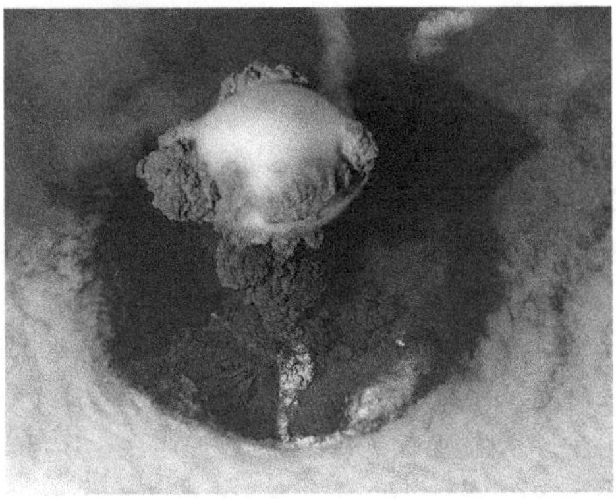

An early stage of the eruption of Sarychev volcano, seen from space.

An explosive eruption always begins with some form of blockage in the crater of a volcano that prevents the release of gases trapped in highly viscous andesitic or rhyolitic magma. The high viscosity of these forms of magma prevents the release of trapped gases. The pressure of the flowing magma builds until eventually the blockage is blasted out in an explosive eruption. The pressure from the magma and gases are released through the weakest point in the cone, usually the crater. However, in the case of the eruption of Mount St. Helens, pressure was released on the side of the volcano, rather than the crater.

The sudden release of pressure causes the gases in the magma to suddenly froth and create volcanic ash and pumice, which is then ejected through the volcanic vent to create the signature eruption column commonly associated with explosive eruptions. The size and duration of the column depends on the volume of magma being released and how much pressure the magma was under.

Types of Explosive Eruption

- Vulcanian eruption,
- Peléan eruption,
- Plinian eruption,
- Phreatomagmatic eruption and Phreatic eruption,
- Surtseyan eruption.

Consequences:

- Eruption column;
- Pyroclastic flow;
- Pyroclastic fall;
- Pyroclastic surge.

Pyroclastic Flows

Pyroclastic flows occur towards the end of an explosive eruption, as pressure begins to decline. The eruption column of ash is supported by pressure from the gases being released, and as the gases are depleted, pressure falls and the eruption column begins to collapse. When the column collapses in on itself, ash and rock fall back down to the ground and begin to flow down the slopes of the volcano. These flows can travel at up to 80 km per hour, and reach temperatures of 200° to 700° Celsius. The high temperatures can cause combustion of any flammable materials in its path, including wood, vegetation, and buildings. When snow and ice melt as a part of an eruption, large amounts of water mixed in with the flow can create lahars. The risk of lahars is particularly high on volcanoes such as Mount Rainier near Seattle and Tacoma, Washington.

Supervolcanoes

The eruptions of supervolcanoes are the rarest of volcanic eruptions but also the most destructive. The timescale between these eruptions is generally marked by hundreds or thousands of years. This type of eruption generally causes destruction on a continental scale, and can also result in the lowering of temperatures worldwide.

Plinian Eruption

Plinian eruptions or Vesuvian eruptions are volcanic eruptions marked by their similarity to the eruption of Mount Vesuvius in 79 AD, which destroyed the ancient Roman cities of Herculaneum and Pompeii. The eruption was described in a letter written by Pliny the Younger, after the death of his uncle Pliny the Elder.

Plinian/Vesuvian eruptions are marked by columns of volcanic debris and hot gases ejected high into the stratosphere, the second layer of Earth's atmosphere. The key characteristics are ejection of large amount of pumice and very powerful continuous gas-driven eruptions. According to the Volcanic Explosivity Index, Plinian eruptions have a VEI of 4, 5 or 6, sub-Plinian 3 or 4, and ultra-Plinian 6, 7 or 8.

A Stone Pine, the type of tree used by Pliny to describe the eruption.

Short eruptions can end in less than a day, but longer events can continue for several days or months. The longer eruptions begin with production of clouds of volcanic ash, sometimes with pyroclastic surges. The amount of magma erupted can be so large that it depletes the magma chamber below, causing the top of the volcano to collapse, resulting in a caldera. Fine ash and pulverized pumice can deposit over large areas. Plinian eruptions are often accompanied by loud noises, such as those generated by the 1883 eruption of Krakatoa. The sudden discharge of electrical charges accumulated in the air around the ascending column of volcanic ashes also often causes lightning strikes as depicted by the English geologist George Julius Poulett Scrope in his painting of 1822.

The lava is usually dacitic or rhyolitic, rich in silica. Basaltic, low-silica lavas are unusual for Plinian eruptions; the most recent basaltic example is the 1886 eruption of Mount Tarawera on New Zealand's North Island.

Eruption cloud from Redoubt Volcano as viewed to the west from the Kenai Peninsula (more than 60 km from the volcano's summit).

Pliny described his uncle's involvement from the first observation of the eruption:

"On August 24th, about one in the afternoon, my mother desired him to observe a cloud which appeared of a very unusual size and shape. He had just taken a turn in the sun and, after bathing himself in cold water, and making a light luncheon, gone back to his books: he immediately arose and went out upon a rising ground from whence he might get a better sight of this very uncommon appearance. A cloud, from which mountain was uncertain, at this distance (but it was found afterwards to come from Mount Vesuvius), was ascending, the appearance of which I cannot give you a more exact description of than by likening it to that of a pine tree, for it shot up to a great height in the form of a very tall trunk, which spread itself out at the top into a sort of branches; occasioned, I imagine, either by a sudden gust of air that impelled it, the force of which decreased as it advanced upwards, or the cloud itself being pressed back again by its own weight, expanded in the manner I have mentioned; it appeared sometimes bright and sometimes dark and spotted, according as it was either more or less impregnated with earth and cinders. This phenomenon seemed to a man of such learning and research as my uncle extraordinary and worth further looking into."

Pliny the Elder set out to rescue the victims from their perilous position on the shore of the Bay of Naples, and launched his galleys, crossing the bay to Stabiae (near the modern town of Castellammare di Stabia). Pliny the Younger provided an account of his death, and suggested that he collapsed and died through inhaling poisonous gases emitted from the volcano. His body was found interred under the ashes of the eruption with no apparent injuries on 26 August, after the plume had dispersed, confirming asphyxiation or poisoning.

Examples of Plinian Eruption

- The Long Valley Caldera eruption in Eastern California, United States, which happened over 760,000 years ago.
- The 4860 BC eruption forming Crater Lake in Oregon, United States.
- The 1645 BC eruption of Thera in the south Aegean Sea, Greece.

- The 400s BC eruption of the Bridge River Vent in British Columbia, Canada.
- The 79 AD eruption of Mount Vesuvius in Pompeii, Italy. It was the prototypical Plinian eruption.
- The 180 AD Lake Taupo eruption in New Zealand.
- The 946 eruption of Paektu Mountain in China/North Korea.
- The 1257 eruption of Mount Samalas in Lombok, Indonesia.
- The 1600 eruption of Huaynaputina in Peru.
- The 1667 and 1739 eruptions of Mount Tarumae in Hokkaido, Japan.
- The 1707 eruption of Mount Fuji in Japan.
- The 1815 eruption of Mount Tambora in the island of Sumbawa, Indonesia.
- The 1883 eruption of Krakatoa in Sunda Strait, Indonesia.
- The 1886 eruption of Mount Tarawera in New Zealand.
- The 1980 eruption of Mount St. Helens in Washington in the United States.
- The 1982 eruption of El Chichón in Chiapanecan Volcanic Arc, Chiapas, Mexico.
- The 1991 eruption of Mount Pinatubo in Zambales, Central Luzon, Philippines.
- The June 2009 eruption of Sarychev Peak in Russia.

Ultra-Plinian

According to the Volcanic Explosivity Index, a VEI of 6 to 8 is classified as "ultra-Plinian". Eruptions of this type are defined by ash plumes over 25 km (16 mi) high and a volume of erupted material 10 km^3 (2 cu mi) to 1,000 km^3 (200 cu mi) in size. Eruptions in the ultra-Plinian category include the Lava Creek eruption of the Yellowstone Caldera (c. 640,000 years ago, VEI 8), Lake Toba (c. 74,000 years ago, VEI 8), Tambora (1815, VEI 7), Krakatoa (1883, VEI 6), Akahoya eruption of Kikai Caldera, Japan, and the 1991 Mount Pinatubo eruption in the Philippines (VEI 6).

Other Types

Lava Domes

Lava domes form when very viscous, rubbly lava (usually andesite, dacite or rhyolite) is squeezed out of a vent without exploding. The lava piles up into a dome, which may grow by inflating from the inside or by squeezing out lobes of lava (something like toothpaste coming out of a tube). These lava lobes can be short and blobby, long and thin, or even form spikes that rise tens of meters into the air before they fall over. Lava domes may be rounded, pancake-shaped, or irregular piles of rock, depending on the type of lava they form from.

In figure, lava dome. Lava domes, such as this example in the crater of Mount St. Helens, are piles of viscous lava that are too cool and sticky to flow far. Domes grow and collapse in cycles, and often form at volcanoes that also experience Plinian eruptions. Photo by Lyn Topinka, USGS, August 12, 1985.

Lava domes are not just passive piles of rock; they can sometimes collapse and form pyroclastic density currents, extrude lava flows, or experience small and large explosive eruptions (which may even destroy the domes!) A dome-building eruption may go on for months or years, but they are usually repetitive (meaning that a volcano will build and destroy several domes before the eruption ceases). Redoubt volcano in Alaska and Chaiten in Chile are currently active examples of this type of eruption, and Mount St. Helens in the state of Washington spent several years building several lava domes.

Surtseyan Eruption

In figure, Surtseyan eruption. Lava erupting through water creates the dramatic plumes of scoria and billowing ash-and-gas clouds of a Surtseyan eruption. The type example of this eruption occurred at Surtsey, a volcanic island off the coast of Iceland. NOAA image of the 1963 eruption.

Surtseyan Eruption

Surtseyan eruptions are a kind of hydromagmatic eruption, where magma or lava interacts explosively with water. In most cases, Surtseyan eruptions occur when an undersea volcano has finally grown large enough to break the water's surface; because water expands when it turns to steam, water that comes into contact with hot lava explodes and creates plumes of ash, steam and scoria. Lavas created by a Surtseyan eruption tend to be basalt, since most oceanic volcanoes are basaltic.

The classic example of a Surtseyan eruption was the volcanic island of Surtsey, which erupted off the south coast of Iceland between 1963 and 1965. Hydromagmatic activity built up several square kilometers of tephra over the first several months of the eruption; eventually, seawater could no longer reach the vent, and the eruption transitioned to Hawaiian and Strombolian styles. More recently, in March 2009, several vents of the volcanic island of Hunga Ha'apai near Tonga began to erupt. The onshore and offshore explosions created plumes of ash and steam that rose to more than 8 km (5 miles) altitude, and threw plumes of tephra hundreds of meters from the vents.

References

- Wang, Amy (20 May 2018). "Lava spatter hits Hawaii man and shatters his leg in first known injury from Kilauea volcano". Washington Post. Retrieved 22 May 2018
- How-do-volcanoes-erupt?Qt-news-science-products=0#qt-news-science-products, faqs: usgs.gov, Retrieved 10 February, 2019
- Burton, Mike; Allard, Patrick; Muré, Filippo; Spina, Alessandro La (13 July 2007). "Magmatic Gas Composition Reveals the Source Depth of Slug-Driven Strombolian Explosive Activity". Science. 317 (5835): 227–230. Doi:10.1126/science.1141900. PMID 17626881
- Types-of-volcanic-eruptions, volcanoes: geology.com, Retrieved 11 March, 2019
- Magnusson, E., M. T. Gudmundsson, M. J. Roberts, G. Sigurosson, F. Hoskuldsson, and B. Oddsson. 2012. Ice-volcano interactions during the 2010 Eyjafjallajökull Eruption, as revealed by airborne imaging radar. Journal of Geophysical Research: Solid Earth, 117, B07405

5
Plate Tectonics and Volcanism

The theory which describes the large-scale movement of the different plates in the Earth's lithosphere is termed as plate tectonics. Volcanism refers to the eruption of lava and volcanic gases through a rupture onto the Earth's surface. This chapter closely examines the key concepts of plate tectonics and volcanism.

Plate tectonics is the theory dealing with the dynamics of Earth's outer shell—the lithosphere—that revolutionized Earth sciences by providing a uniform context for understanding mountain-building processes, volcanoes, and earthquakes as well as the evolution of Earth's surface and reconstructing its past continents and oceans.

The concept of plate tectonics was formulated in the 1960s. According to the theory, Earth has a rigid outer layer, known as the lithosphere, which is typically about 100 km (60 miles) thick and overlies a plastic (moldable, partially molten) layer called the asthenosphere. The lithosphere is broken up into seven very large continental- and ocean-sized plates, six or seven medium-sized regional plates, and several small ones. These plates move relative to each other, typically at rates of 5 to 10 cm (2 to 4 inches) per year, and interact along their boundaries, where they converge, diverge, or slip past one another. Such interactions are thought to be responsible for most of Earth's seismic and volcanic activity, although earthquakes and volcanoes can occur in plate interiors. Plate motions cause mountains to rise where plates push together, or converge, and continents to fracture and oceans to form where plates pull apart, or diverge. The continents are embedded in the plates and drift passively with them, which over millions of years results in significant changes in Earth's geography.

The theory of plate tectonics is based on a broad synthesis of geologic and geophysical data. It is now almost universally accepted, and its adoption represents a true scientific revolution, analogous in its consequences to quantum mechanics in physics or the discovery of the genetic code in biology. Incorporating the much older idea of continental drift, as well as the concept of seafloor spreading, the theory of plate tectonics has provided an overarching framework in which to describe the past geography of continents and oceans, the processes controlling creation and destruction of landforms, and the evolution of Earth's crust, atmosphere, biosphere, hydrosphere, and climates. During the late 20th and early 21st centuries, it became apparent that plate-tectonic processes profoundly influence the composition of Earth's atmosphere and oceans, serve as a prime cause of long-term climate change, and make significant contributions to the chemical and physical environment in which life evolves.

Principles of Plate Tectonics

In essence, plate-tectonic theory is elegantly simple. Earth's surface layer, 50 to 100 km (30 to 60 miles) thick, is rigid and is composed of a set of large and small plates. Together, these plates constitute the lithosphere, The lithosphere rests on and slides over an underlying partially molten (and thus weaker but generally denser) layer of plastic partially molten rock known as the asthenosphere, Plate movement is possible because the lithosphere-asthenosphere boundary is a zone of detachment. As the lithospheric plates move across Earth's surface, driven by forces as yet not fully understood, they interact along their boundaries, diverging, converging, or slipping past each other. While the interiors of the plates are presumed to remain essentially undeformed, plate boundaries are the sites of many of the principal processes that shape the terrestrial surface, including earthquakes, volcanism, and orogeny (that is, formation of mountain ranges).

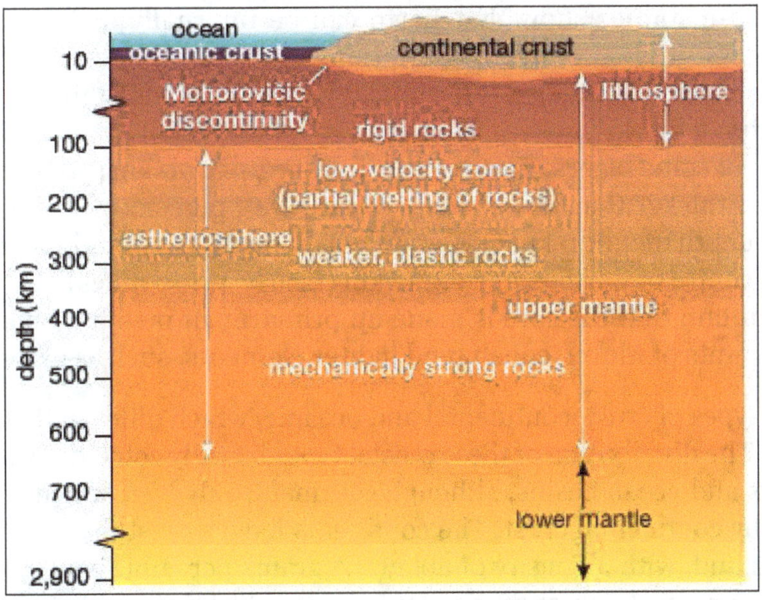

A cross section of Earth's outer layers, from the crust through the lower mantle.

The process of plate tectonics may be driven by convection in Earth's mantle, the pull of heavy old pieces of crust into the mantle, or some combination of both. For a deeper discussion of plate-driving mechanisms, see Plate-driving mechanisms and the role of the mantle.

Earth's Layers

Knowledge of Earth's interior is derived primarily from analysis of the seismic waves that propagate through Earth as a result of earthquakes. Depending on the material they travel through, the waves may either speed up, slow down, bend, or even stop if they cannot penetrate the material they encounter.

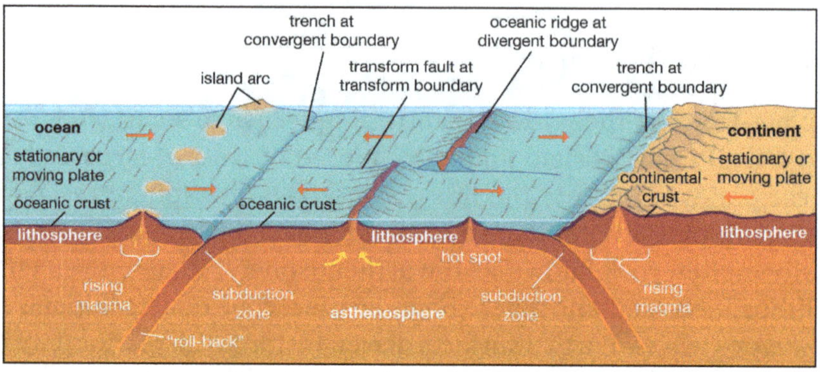

Crustal generation and destruction.

Three-dimensional diagram showing crustal generation and destruction according to the theory of plate tectonics; included are the three kinds of plate boundaries—divergent, convergent (or collision), and strike-slip (or transform).

Collectively, these studies show that Earth can be internally divided into layers on the basis of either gradual or abrupt variations in chemical and physical properties. Chemically, Earth can be divided into three layers. A relatively thin crust, which typically varies from a few kilometres to 40 km (about 25 miles) in thickness, sits on top of the mantle. (In some places, Earth's crust may be up to 70 km [40 miles] thick.) The mantle is much thicker than the crust; it contains 83 percent of Earth's volume and continues to a depth of 2,900 km (1,800 miles). Beneath the mantle is the core, which extends to the centre of Earth, some 6,370 km (nearly 4,000 miles) below the surface. Geologists maintain that the core is made up primarily of metallic iron accompanied by smaller amounts of nickel, cobalt, and lighter elements, such as carbon and sulfur.

There are two types of crust, continental and oceanic, which differ in their composition and thickness. The distribution of these crustal types broadly coincides with the division into continents and ocean basins, although continental shelves, which are submerged, are underlain by continental crust. The continents have a crust that is broadly granitic in composition and, with a density of about 2.7 grams per cubic cm (0.098 pound per cubic inch), is somewhat lighter than oceanic crust, which is basaltic (i.e., richer in iron and magnesium than granite) in composition and has a density of about 2.9 to 3 grams per cubic cm (0.1 to 0.11 pound per cubic inch). Continental crust is typically 40 km (25 miles) thick, while oceanic crust is much thinner, averaging about 6 km (4 miles) in thickness. These crustal rocks both sit on top of the mantle, which is ultramafic in composition (i.e., very rich in magnesium and iron-bearing silicate minerals). The boundary between the crust (continental or oceanic) and the underlying mantle is known as the Mohorovičić discontinuity (also called Moho), which is named for its discoverer, Croatian seismologist Andrija Mohorovičić. The Moho is clearly defined by seismic studies, which detect an acceleration in seismic waves as they pass from the crust into the denser mantle. The boundary between the mantle and the core is also clearly defined by seismic studies, which suggest that the outer part of the core is a liquid.

The effect of the different densities of lithospheric rock can be seen in the different average elevations of continental and oceanic crust. The less-dense continental crust has greater buoyancy, causing it to float much higher in the mantle. Its average elevation above sea level is 840 metres (2,750 feet), while the average depth of oceanic crust is 3,790 metres (12,400 feet). This density difference creates two principal levels of Earth's surface.

The lithosphere itself includes all the crust as well as the upper part of the mantle (i.e., the region directly beneath the Moho), which is also rigid. However, as temperatures increase with depth, the heat causes mantle rocks to lose their rigidity. This process begins at about 100 km (60 miles) below the surface. This change occurs within the mantle and defines the base of the lithosphere and the top of the asthenosphere. This upper portion of the mantle, which is known as the lithospheric mantle, has an average density of about 3.3 grams per cubic cm (0.12 pound per cubic inch). The asthenosphere, which sits directly below the lithospheric mantle, is thought to be slightly denser at 3.4–4.4 grams per cubic cm (0.12–0.16 pound per cubic inch).

In contrast, the rocks in the asthenosphere are weaker, because they are close to their melting temperatures. As a result, seismic waves slow as they enter the asthenosphere. With increasing depth, however, the greater pressure from the weight of the rocks above causes the mantle to become gradually stronger, and seismic waves increase in velocity, a defining characteristic of the lower mantle. The lower mantle is more or less solid, but the region is also very hot, and thus the rocks can flow very slowly (a process known as creep).

During the late 20th and early 21st centuries, scientific understanding of the deep mantle was greatly enhanced by high-resolution seismological studies combined with numerical modeling and laboratory experiments that mimicked conditions near the core-mantle boundary. Collectively, these studies revealed that the deep mantle is highly heterogeneous and that the layer may play a fundamental role in driving Earth's plates.

At a depth of about 2,900 km (1,800 miles), the lower mantle gives way to Earth's outer core, which is made up of a liquid rich in iron and nickel. At a depth of about 5,100 km (3,200 miles), the outer core transitions to the inner core. Although it has a higher temperature than the outer core, the inner core is solid because of the tremendous pressures that exist near Earth's centre. Earth's inner core is divided into the outer-inner core (OIC) and the inner-inner core (IIC), which differ from one another with respect to the polarity of their iron crystals. The polarity of the iron crystals of the OIC is oriented in a north-south direction, whereas that of the IIC is oriented east-west.

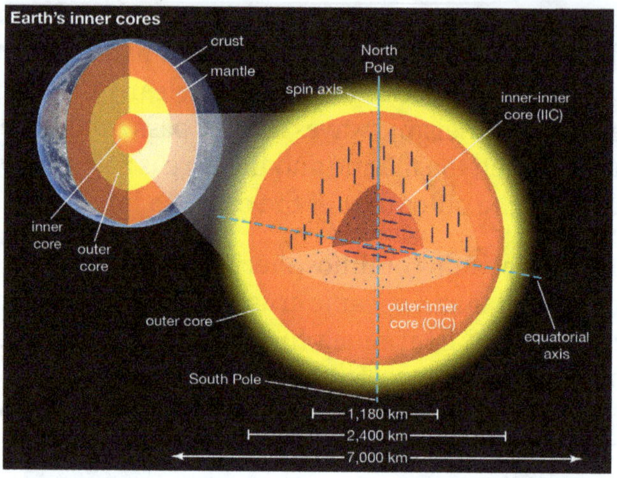

The internal layers of Earth's core, including its two inner cores.

Plate Boundaries

Lithospheric plates are much thicker than oceanic or continental crust. Their boundaries do not usually coincide with those between oceans and continents, and their behaviour is only partly influenced by whether they carry oceans, continents, or both. The Pacific Plate, for example, is entirely oceanic, whereas the North American Plate is capped by continental crust in the west (the North American continent) and by oceanic crust in the east and extends under the Atlantic Ocean as far as the Mid-Atlantic Ridge.

A detai movement of plate A to the left relative to plates B and C results in several types of simultaneous interactions along the plate boundaries. At the rear, plates A and B move apart, or diverge, resulting in extension and the formation of a divergent margin. At the front, plates A and B overlap, or converge, resulting in compression and the formation of a convergent margin. Along the sides, the plates slide past one another, a process called shear. As these zones of shear link other plate boundaries to one another, they are called transform faults.

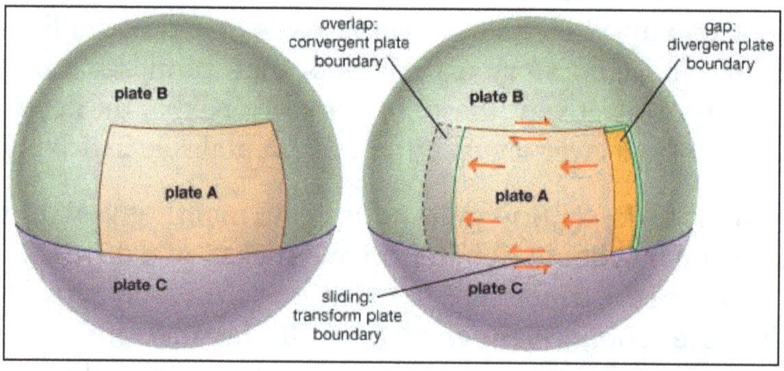

Theoretical diagram showing the effects of an advancing tectonic plate on other adjacent, but stationary, tectonic plates. At the advancing edge of plate A, the overlap with plate B creates a convergent boundary. In contrast, the gap left behind the trailing edge of plate A forms a divergent boundary with plate B. As plate A slides past portions of both plate B and plate C, transform boundaries develop.

Divergent Margins

As plates move apart at a divergent plate boundary, the release of pressure produces partial melting of the underlying mantle. This molten material, known as magma, is basaltic in composition and is buoyant. As a result, it wells up from below and cools close to the surface to generate new crust. Because new crust is formed, divergent margins are also called constructive margins.

Continental Rifting

Upwelling of magma causes the overlying lithosphere to uplift and stretch. (Whether magmatism [the formation of igneous rock from magma] initiates the rifting or whether rifting decompresses the mantle and initiates magmatism is a matter of significant debate.) If the diverging plates are capped by continental crust, fractures develop that are invaded by the ascending magma, prying the continents farther apart. Settling of the continental blocks creates a rift valley, such as the present-day East African Rift Valley. As the rift continues to widen, the continental crust becomes progressively thinner until separation of the plates is achieved and a new ocean is created. The ascending partial melt cools and crystallizes to form new crust. Because the partial melt is basaltic in composition, the new crust is oceanic, and an ocean ridge develops along the site of the former continental rift. Consequently, diverging plate boundaries, even if they originate within continents, eventually come to lie in ocean basins of their own making.

Rift valley; Thingvellir National Park.

The Thingvellir fracture zone at Thingvellir National Park in southwestern Iceland is an example of a rift valley. The Thingvellir fracture lies in the Mid-Atlantic Ridge, which extends through the centre of Iceland.

Seafloor Spreading

As upwelling of magma continues, the plates continue to diverge, a process known as seafloor spreading. Samples collected from the ocean floor show that the age of oceanic crust increases with distance from the spreading centre—important evidence in favour of this process. These age data also allow the rate of seafloor spreading to be determined, and they show that rates vary from about 0.1 cm (0.04 inch) per year to 17 cm (6.7 inches) per year. Seafloor-spreading rates are much more rapid in the Pacific Ocean than in the Atlantic and Indian oceans. At spreading rates of about 15 cm (6 inches) per year, the entire crust beneath the Pacific Ocean (about 15,000 km [9,300 miles] wide) could be produced in 100 million years.

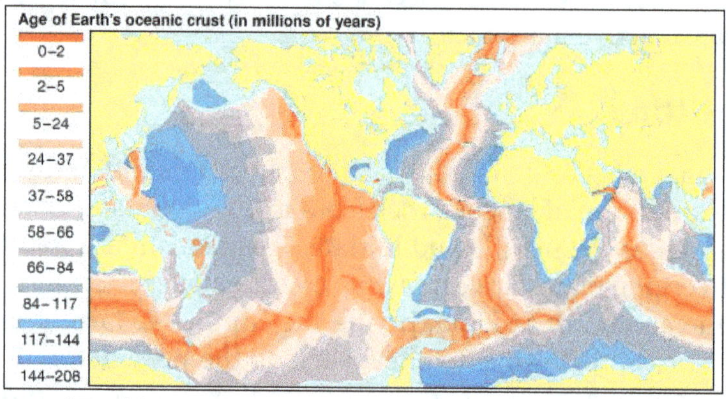

Map showing the age of Earth's oceanic crust and the pattern of seafloor spreading at the global scale.

Divergence and creation of oceanic crust are accompanied by much volcanic activity and by many shallow earthquakes as the crust repeatedly rifts, heals, and rifts again. Brittle earthquake-prone rocks occur only in the shallow crust. Deep earthquakes, in contrast, occur less frequently, due to the high heat flow in the mantle rock. These regions of oceanic crust are swollen with heat and so are elevated by 2 to 3 km (1.2 to 1.9 miles) above the surrounding seafloor. The elevated topography results in a feedback scenario in which the resulting gravitational force pushes the crust apart, allowing new magma to well up from below, which in turn sustains the elevated topography. Its summits are typically 1 to 5 km (0.6 to 3.1 miles) below the ocean surface. On a global scale, these ridges form an interconnected system of undersea "mountains" that are about 65,000 km (40,000 miles) in length and are called oceanic ridges.

Convergent Margins

Given that Earth is constant in volume, the continuous formation of Earth's new crust produces an excess that must be balanced by destruction of crust elsewhere. This is accomplished at convergent plate boundaries, also known as destructive plate boundaries, where one plate descends at an angle—that is, is subducted—beneath the other.

Because oceanic crust cools as it ages, it eventually becomes denser than the underlying asthenosphere, and so it has a tendency to subduct, or dive under, adjacent continental plates or younger sections of oceanic crust. The life span of the oceanic crust is prolonged by its rigidity, but eventually this resistance is overcome. Experiments show that the subducted oceanic lithosphere is denser than the surrounding mantle to a depth of at least 600 km (about 400 miles).

The mechanisms responsible for initiating subduction zones are controversial. During the late 20th and early 21st centuries, evidence emerged supporting the notion that subduction zones preferentially initiate along preexisting fractures (such as transform faults) in the oceanic crust. Irrespective of the exact mechanism, the geologic record indicates that the resistance to subduction is overcome eventually.

Where two oceanic plates meet, the older, denser plate is preferentially subducted beneath the younger, warmer one. Where one of the plate margins is oceanic and the other is continental, the greater buoyancy of continental crust prevents it from sinking, and the oceanic plate is preferentially subducted. Continents are preferentially preserved in this manner relative to oceanic crust, which is continuously recycled into the mantle. This explains why ocean floor rocks are generally less than 200 million years old whereas the oldest continental rocks are more than 4 billion years old. Before the middle of the 20th century, most geoscientists maintained that continental crust was too buoyant to be subducted. However, it later became clear that slivers of continental crust adjacent to the deep-sea trench, as well as sediments deposited in the trench, may be dragged down the subduction zone. The recycling of this material is detected in the chemistry of volcanoes that erupt above the subduction zone.

Two plates carrying continental crust collide when the oceanic lithosphere between them has been eliminated. Eventually, subduction ceases and towering mountain ranges, such as the Himalayas, are created.

Because the plates form an integrated system, it is not necessary that new crust formed at any given divergent boundary be completely compensated at the nearest subduction zone, as long as the total amount of crust generated equals that destroyed.

Subduction Zones

The subduction process involves the descent into the mantle of a slab of cold hydrated oceanic lithosphere about 100 km (60 miles) thick that carries a relatively thin cap of oceanic sediments. The path of descent is defined by numerous earthquakes along a plane that is typically inclined between 30° and 60° into the mantle and is called the Wadati-Benioff zone, for Japanese seismologist Kiyoo Wadati and American seismologist Hugo Benioff, who pioneered its study. Between 10 and 20 percent of the subduction zones that dominate the circum-Pacific ocean basin are subhorizontal (that is, they subduct at angles between 0° and 20°). The factors that govern the dip of the subduction zone are not fully understood, but they probably include the age and thickness of the subducting oceanic lithosphere and the rate of plate convergence.

A subducting plate's path (called the Benioff-Wadati [or Wadati-Benioff] zone) is defined by numerous earthquakes along a plane that is typically inclined between 30° and 60° into the mantle.

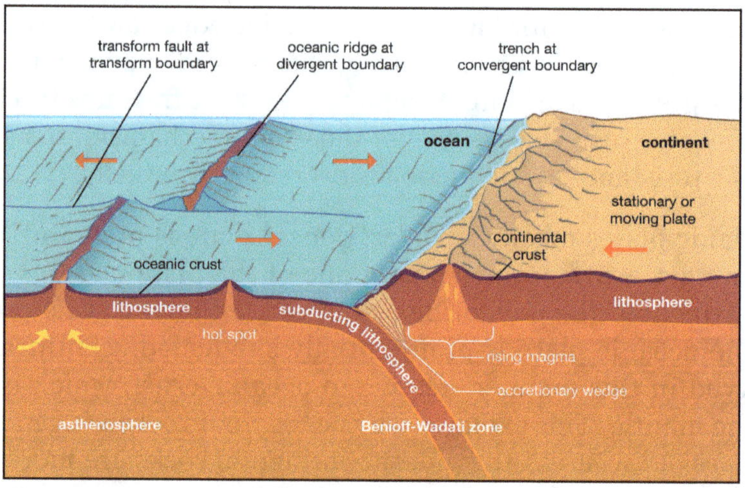

Subducting tectonic plate.

Most, but not all, earthquakes in this planar dipping zone result from compression, and the seismic activity extends 300 to 700 km (200 to 400 miles) below the surface, implying that the subducted crust retains some rigidity to this depth. At greater depths the subducted plate is partially recycled into the mantle.

The site of subduction is marked by a deep trench, between 5 and 11 km (3 and 7 miles) deep, that is produced by frictional drag between the plates as the descending plate bends before it subducts. The overriding plate scrapes sediments and elevated portions of ocean floor off the upper crust of the lower plate, creating a zone of highly deformed rocks within the trench that becomes attached, or accreted, to the overriding plate. This chaotic mixture is known as an accretionary wedge.

The rocks in the subduction zone experience high pressures but relatively low temperatures, an effect of the descent of the cold oceanic slab. Under these conditions the rocks recrystallize, or metamorphose, to form a suite of rocks known as blueschists, named for the diagnostic blue mineral called glaucophane, which is stable only at the high pressures and low temperatures found in subduction zones. At deeper levels in the subduction zone (that is, greater than 30–35 km [about 19–22 miles]), eclogites, which consist of high-pressure minerals such as red garnet (pyrope) and omphacite (pyroxene), form. The formation of eclogite from blueschist is accompanied by a significant increase in density and has been recognized as an important additional factor that facilitates the subduction process.

Island Arcs

When the downward-moving slab reaches a depth of about 100 km (60 miles), it gets sufficiently warm to drive off its most volatile components, thereby stimulating partial melting of mantle in the plate above the subduction zone (known as the mantle wedge).

Melting in the mantle wedge produces magma, which is predominantly basaltic in composition. This magma rises to the surface and gives birth to a line of volcanoes in the overriding plate, known as a volcanic arc, typically a few hundred kilometres behind the oceanic trench. The distance between the trench and the arc, known as the arc-trench gap, depends on the angle of subduction. Steeper subduction zones have relatively narrow arc-trench gaps. A basin may form within this region, known as a fore-arc basin, and may be filled with sediments derived from the volcanic arc or with remains of oceanic crust.

If both plates are oceanic, as in the western Pacific Ocean, the volcanoes form a curved line of islands, known as an island arc, that is parallel to the trench, as in the case of the Mariana Islands and the adjacent Mariana Trench. If one plate is continental, the volcanoes form inland, as they do in the Andes of western South America. Though the process of magma generation is similar, the ascending magma may change its composition as it rises through the thick lid of continental crust, or it may provide sufficient heat to melt the crust. In either case, the composition of the volcanic mountains formed tends to be more silicon-rich and iron- and magnesium-poor relative to the volcanic rocks produced by ocean-ocean convergence.

Back-arc Basins

Where both converging plates are oceanic, the margin of the older oceanic crust will be subducted because older oceanic crust is colder and therefore more dense. As the dense slab collapses into the asthenosphere, however, it also may "roll back" oceanward and cause extension in the overlying plate. This results in a process known as back-arc spreading, in which a basin opens up behind the island arc. The crust behind the arc becomes progressively thinner, and the decompression of the underlying mantle causes the crust to melt, initiating seafloor-spreading processes, such as melting and the production of basalt; these processes are similar to those that occur at ocean ridges. The geochemistry of the basalts produced at back-arc basins superficially resembles that of basalts produced at ocean ridges, but subtle trace element analyses can detect the influence of a nearby subducted slab.

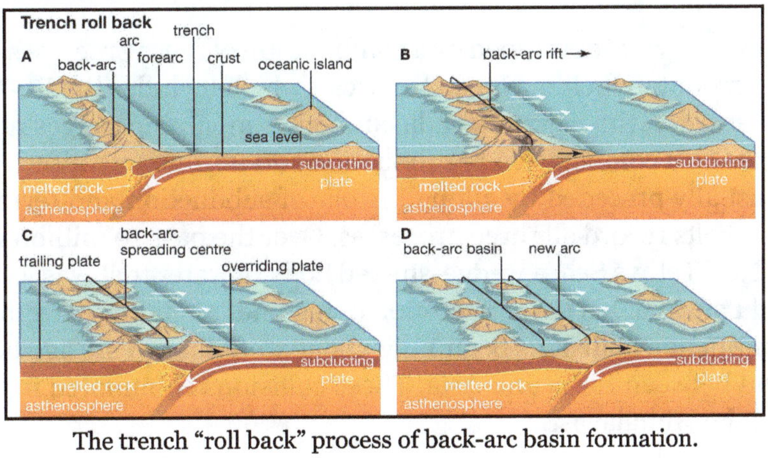

The trench "roll back" process of back-arc basin formation.

This style of subduction predominates in the western Pacific Ocean, in which a number of back-arc basins separate several island arcs from Asia. Examples include the Mariana Islands, the Kuril Islands, and the main islands of Japan. However, if the rate of convergence increases or if anomalously thick oceanic crust (possibly caused by rising mantle plume activity) is conveyed into the subduction zone, the slab may flatten. Such flattening causes the back-arc basin to close, resulting in deformation, metamorphism, and even melting of the strata deposited in the basin.

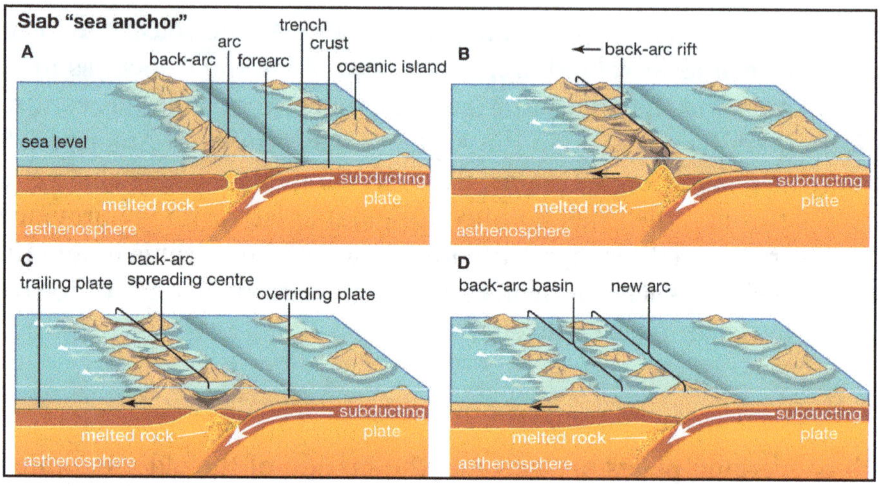

The slab "sea anchor" process of back-arc basin formation.

Mountain Building

If the rate of subduction in an ocean basin exceeds the rate at which the crust is formed at oceanic ridges, a convergent margin forms as the ocean initially contracts. This process can lead to collision between the approaching continents, which eventually terminates subduction. Mountain building can occur in a number of ways at a convergent margin: mountains may rise as a consequence of the subduction process itself, by the accretion of small crustal fragments (which, along with linear island chains and oceanic ridges, are known as terranes), or by the collision of two large continents.

Many mountain belts were developed by a combination of these processes. For example, the Cordilleran mountain belt of North America—which includes the Rocky Mountains as well as the Cascades, the Sierra Nevada, and other mountain ranges near the Pacific coast—developed by a combination of subduction and terrane accretion. As continental collisions are usually preceded by a long history of subduction and terrane accretion, many mountain belts record all three processes. Over the past 70 million years the subduction of the Neo-Tethys Sea, a wedge-shaped body of water that was located between Gondwana and Laurasia, led to the accretion of terranes along the margins of Laurasia, followed by continental collisions beginning about 30 million years ago between Africa and Europe and between India and Asia. These collisions culminated in the formation of the Alps and the Himalayas.

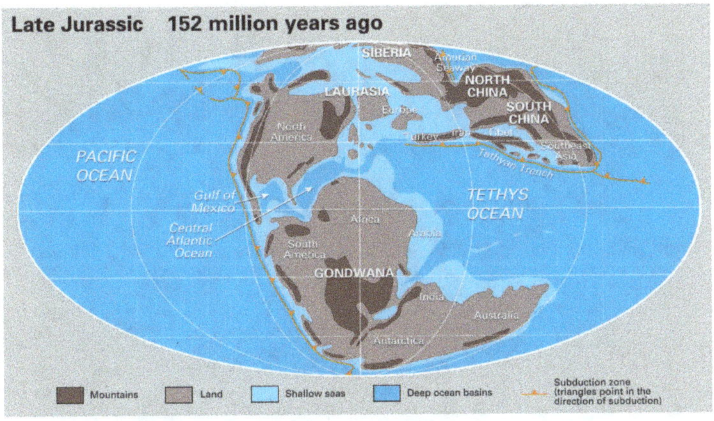

Jurassic paleogeography.

Distribution of landmasses, mountainous regions, shallow seas, and deep ocean basins during the late Jurassic Period. Included in the paleogeographic reconstruction are the locations of the interval's subduction zones.

Mountains by Subduction

Mountain building by subduction is classically demonstrated in the Andes Mountains of South America. Subduction results in voluminous magmatism in the mantle and crust overlying the subduction zone, and, therefore, the rocks in this region are warm and weak. Although subduction is a long-term process, the uplift that results in mountains tends to occur in discrete episodes and may reflect intervals of stronger plate convergence that squeezes the thermally weakened crust upward. For example, rapid uplift of the Andes approximately 25 million years ago is evidenced by a reversal in the flow of the Amazon River from its ancestral path toward the Pacific Ocean to its modern path, which empties into the Atlantic Ocean.

In addition, models have indicated that the episodic opening and closing of back-arc basins have been the major factors in mountain-building processes, which have influenced the plate-tectonic evolution of the western Pacific for at least the past 500 million years.

Mountains by Terrane Accretion

As the ocean contracts by subduction, elevated regions within the ocean basin—terranes—are transported toward the subduction zone, where they are scraped off the descending plate and added—accreted—to the continental margin. Since the late Devonian and early Carboniferous periods, some 360 million years ago, subduction beneath the western margin of North America has resulted in several collisions with terranes. The piecemeal addition of these accreted terranes has added an average of 600 km (400 miles) in width along the western margin of the North American continent, and the collisions have resulted in important pulses of mountain building.

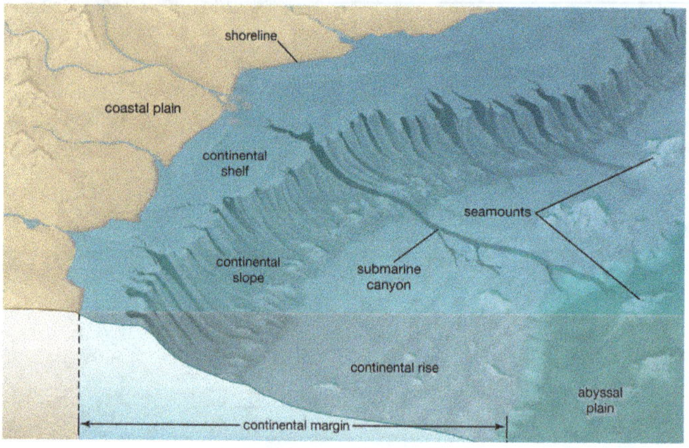

Continental margin .

The broad, gentle pitch of the continental shelf gives way to the relatively steep continental slope. The more gradual transition to the abyssal plain is a sediment-filled region called the continental rise. The continental shelf, slope, and rise are collectively called the continental margin.

During these accretionary events, small sections of the oceanic crust may break away from the subducting slab as it descends. Instead of being subducted, these slices are thrust over the overriding plate and are said to be obducted. Where this occurs, rare slices of ocean crust, known as ophiolites, are preserved on land. They provide a valuable natural laboratory for studying the composition and character of the oceanic crust and the mechanisms of their emplacement and preservation on land. A classic example is the Coast Range ophiolite of California, which is one of the most extensive ophiolite terranes in North America. These ophiolite deposits run from the Klamath Mountains in northern California southward to the Diablo Range in central California. This oceanic crust likely formed during the middle of the Jurassic Period, roughly 170 million years ago, in an extensional regime within either a back-arc or a forearc basin. In the late Mesozoic, it was accreted to the western North American continental margin.

Because preservation of oceanic crust is rare, the recognition of ophiolite complexes is very important in tectonic analyses. Until the mid-1980s, ophiolites were thought to represent vestiges of the main oceanic tract, but geochemical analyses have clearly indicated that most ophiolites form near volcanic arcs, such as in back-arc basins characterized by subduction roll-back (the collapse of the subducting plate that causes the extension of the overlying plate). The recognition of ophiolite complexes is very important in tectonic analysis, because they provide insights into the generation of magmatism in oceanic domains, as well as their complex relationships with subduction processes.

Mountains by Continental Collision

Continental collision involves the forced convergence of two buoyant plate margins that results in neither continent being subducted to any appreciable extent. A complex sequence of events ensues that compels one continent to override the other. These processes result in crustal thickening and intense deformation that forces the crust skyward to form huge mountains with crustal roots that extend as deep as 80 km (about 50 miles) relative to Earth's surface, in accordance with the principles of isostasy.

The subducted slab still has a tendency to sink and may become detached and founder (submerge) into the mantle. The crustal root undergoes metamorphic reactions that result in a significant increase in density and may cause the root to also founder into the mantle. Both processes result in a significant injection of heat from the compensatory upwelling of asthenosphere, which is an important contribution to the rise of the mountains.

Continental collisions produce lofty landlocked mountain ranges such as the Himalayas. Much later, after these ranges have been largely leveled by erosion, it is possible that the original contact, or suture, may be exposed.

The balance between creation and destruction on a global scale is demonstrated by the expansion of the Atlantic Ocean by seafloor spreading over the past 200 million years, compensated by the contraction of the Pacific Ocean, and the consumption of an entire ocean between India and Asia (the Tethys Sea). The northward migration of India led to collision with Asia some 40 million years ago. Since that time India has advanced a further 2,000 km (1,250 miles) beneath Asia, pushing up the Himalayas and forming the Plateau of Tibet. Pinned against stable Siberia, China and Indochina were pushed sideways, resulting in strong seismic activity thousands of kilometres from the site of the continental collision.

Transform Faults

Section of the San Andreas Fault in the Carrizo Plain, western California.

Along the third type of plate boundary, two plates move laterally and pass each other along giant fractures in Earth's crust. Transform faults are so named because they are linked to other types of plate boundaries. The majority of transform faults link the offset segments of oceanic ridges. However, transform faults also occur between plate margins with continental crust—for example, the San Andreas Fault in California and the North Anatolian fault system in Turkey. These boundaries are conservative because plate interaction occurs without creating or destroying crust. Because the only motion along these faults is the sliding of plates past each other, the horizontal direction along the fault surface must parallel the direction of plate motion. The fault surfaces are rarely smooth, and pressure may build up when the plates on either side temporarily lock. This buildup of stress may be suddenly released in the form of an earthquake.

Many transform faults in the Atlantic Ocean are the continuation of major faults in adjacent continents, which suggests that the orientation of these faults might be inherited from preexisting weaknesses in continental crust during the earliest stages of the development of oceanic crust. On the other hand, transform faults may themselves be reactivated, and recent geodynamic models suggest that they are favourable environments for the initiation of subduction zones.

Plate motion

Euler's Contributions

In the 18th century, Swiss mathematician Leonhard Euler showed that the movement of a rigid body across the surface of a sphere can be described as a rotation (or turning) around an axisthat goes through the centre of the sphere, known as the axis of rotation. The location of this axis bears no relationship to Earth's spin axis. The point of emergence of the axis through the surface of the sphere is known as the pole of rotation. This theorem of spherical geometryprovides an elegant way to define the motion of the lithospheric plates across Earth's surface. Therefore, the relative motion of two rigid plates may be described as rotations around a common axis, known as the axis of spreading. Application of the theorem requires that the plates not be internally deformed—a requirement not absolutely adhered to but one that appears to be a reasonable approximation of what actually happens. Application of this theorem permits the mathematical reconstruction of past plate configurations.

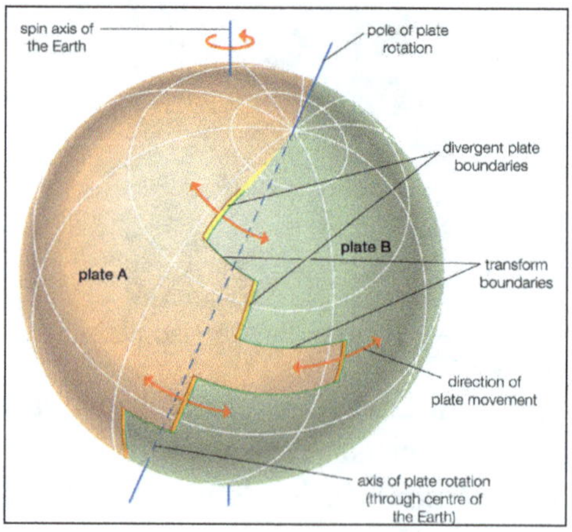

In figure, theoretical depiction of the movement of tectonic plates across Earth's surface.

Movement on a sphere of two plates, A and B, can be described as a rotation around a common pole. Circles around that pole correspond to the orientation of transform faults (that is, single lines in the horizontal that connect to divergent plate boundaries, marked by double lines, in the vertical).

Because all plates form a closed system, all movements can be defined by dealing with them two at a time. The joint pole of rotation of two plates can be determined from their transform boundaries, which are by definition parallel to the direction of motion. Thus, the plates move along transform faults, whose trace defines circles of latitude perpendicular to the axis of spreading, and so form small circles around the pole of rotation. A geometric necessity of this theorem—that lines perpendicular to the transform faults converge on the pole of rotation—is confirmed by measurements. According to this theorem, the rate of plate motion should be slowest near the pole of rotation and increase progressively to a maximum rate along fractures with a 90° angle to it. This relationship is also confirmed by accurate measurements of seafloor-spreading rates.

Leonhard Euler.

Past Plate Movements

Plate tectonics involves the movements of Earth's lithospheric plates relative to one another over the planet's weak asthenosphere. This activity changes the positions of all plates with respect to Earth's spin axis and the Equator. To determine the true geographic positions of the plates in the past, investigators have to define their motions, not only relative to each other but also relative to this independent frame of reference. Hotspots, as classically interpreted, provide an example of such a reference frame, assuming they are the sources of plumes that originate within the deep mantle and have relatively fixed positions over time. If this assumption is valid, the motion of the lithosphere above these plumes can be deduced. The hotspot island chains serve this purpose, their trends providing the direction of motion of a plate. The speed of the plate can be inferred from the increase in age of the volcanoes along the chain relative to the distance between the islands.

Earth scientists are able to accurately reconstruct the positions and movements of plates for the past 150 million to 200 million years because they have the oceanic crust record to provide them with plate speeds and direction of movement. However, since older oceanic crust is continuously consumed to make room for new crust, this kind of evidence is not available for earlier intervals of geologic time, making it necessary for investigators to turn to other, less-precise techniques.

Mantle Convection

Mantle convection is the slow creeping motion of Earth's solid silicate mantle caused by convection currents carrying heat from the interior to the planet's surface.

The Earth's surface lithosphere rides atop the asthenosphere and the two form the components of the upper mantle. The lithosphere is divided into a number of plates that are continuously being created and consumed at their opposite plate boundaries. Accretion occurs as mantle is added to the growing edges of a plate, associated with seafloor spreading. This hot added material cools down by conduction and convection of heat. At the consumption edges of the plate, the material has thermally contracted to become dense, and it sinks under its own weight in the process of subduction usually at an ocean trench.

This subducted material sinks through the Earth's interior. Some subducted material appears to reach the lower mantle, while in other regions, this material is impeded from sinking further, possibly due to a phase transition from spinel to silicate perovskite and magnesiowustite, an endothermic reaction.

The subducted oceanic crust triggers volcanism, although the basic mechanisms are varied. Volcanism may occur due to processes that add buoyancy to partially melted mantle, which would cause upward flow of the partial melt due to decrease in its density. Secondary convection may cause surface volcanism as a consequence of intraplate extension and mantle plumes.

Mantle convection causes tectonic plates to move around the Earth's surface. It seems to have been much more active during the Hadean period, resulting in gravitational sorting of heavier molten iron, nickel, and sulphides to the core and lighter silicate minerals to the mantle.

Types of Convection

During the late 20th century, there was significant debate within the geophysics community as to whether convection is likely to be "layered" or "whole". Although elements of this debate still continue, results from seismic tomography, numerical simulations of mantle convection and examination of Earth's gravitational field are all beginning to suggest the existence of 'whole' mantle convection, at least at the present time. In this model, cold, subducting oceanic lithosphere descends all the way from the surface to the core–mantle boundary (CMB) and hot plumes rise from the CMB all the way to the surface. This picture is strongly based on the results of global seismic tomography models, which typically show slab and plume-like anomalies crossing the mantle transition zone.

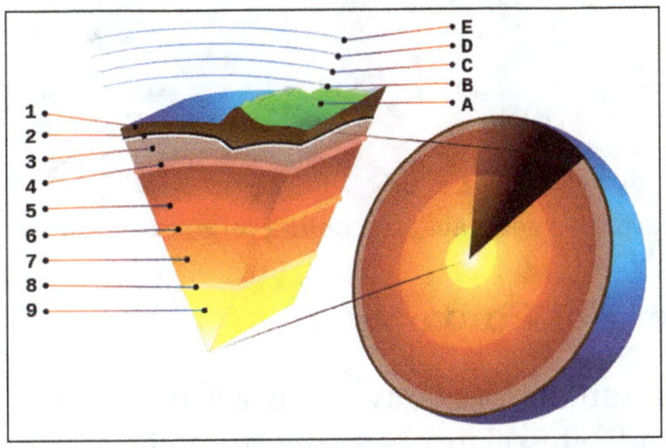

Earth cross-section showing location of upper (3) and lower (5) mantle.

Although it is now well accepted that subducting slabs cross the mantle transition zone and descend into the lower mantle, debate about the existence and continuity of plumes persists, with important implications for the style of mantle convection. This debate is linked to the controversy regarding whether intraplate volcanism is caused by shallow, upper-mantle processes or by plumes from the lower mantle. Many geochemistry studies have argued that the lavas erupted in intraplate areas are different in composition from shallow-derived mid-ocean ridge basalts (MORB). Specifically, they typically have elevated Helium-3 – Helium-4 ratios. Being a primordial nuclide, Helium-3 is not naturally produced on earth. It also quickly escapes from earth's atmosphere when erupted. The elevated He-3/He-4 ratio of Ocean Island Basalts (OIBs) suggest that they must be sources from a part of the earth that has not previously been melted and reprocessed in the same way as MORB source has been. This has been interpreted as their originating from a different, less well-mixed, region, suggested to be the lower mantle. Others, however, have pointed out that geochemical differences could indicate the inclusion of a small component of near-surface material from the lithosphere.

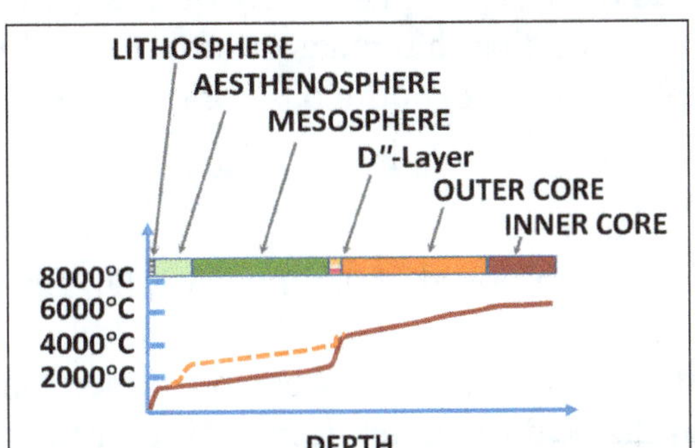

Earth's temperature vs depth. Dashed curve: layered mantle convection. Solid curve: whole-mantle convection.

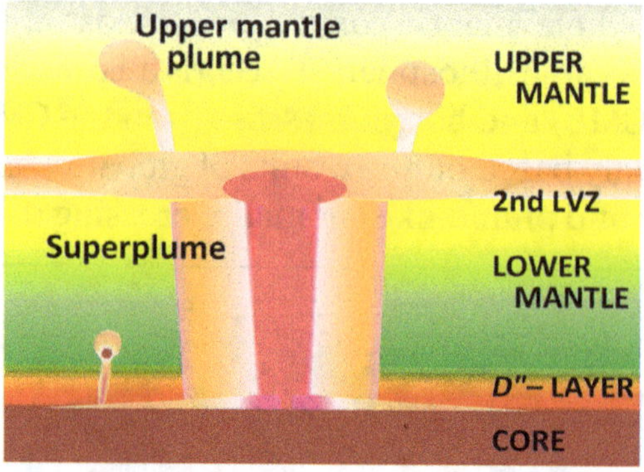

A superplume generated by cooling processes in the mantle.

Planform and Vigour of Convection

On Earth, the Rayleigh number for convection within Earth's mantle is estimated to be of order 10^7, which indicates vigorous convection. This value corresponds to whole mantle convection (i.e. convection extending from the Earth's surface to the border with the core). On a global scale, surface expression of this convection is the tectonic plate motions, and therefore has speeds of a few cm/a. Speeds can be faster for small-scale convection occurring in low-viscosity regions beneath the lithosphere, and slower in the lowermost mantle where viscosities are larger. A single shallow convection cycle takes on the order of 50 million years, though deeper convection can be closer to 200 million years.

Currently, whole mantle convection is thought to include broad-scale downwelling beneath the Americas and the Western Pacific, both regions with a long history of subduction, and upwelling flow beneath the central Pacific and Africa, both of which exhibit dynamic topography consistent with upwelling. This broad-scale pattern of flow is also consistent with the tectonic plate motions, which are the surface expression of convection in the Earth's mantle and currently indicate degree-2 convergence toward the western Pacific and the Americas, and divergence away from the central Pacific and Africa. The persistence of net tectonic divergence away from Africa and the Pacific for the past 250 Myr indicates the long-term stability of this general mantle flow pattern, and is consistent with other studies that suggest long-term stability of the LLSVP regions of the lowermost mantle that form the base of these upwellings.

Creep in the Mantle

Since the mantle is primarily composed of olivine ($(Mg,Fe)_2SiO_4$), the rheological characteristics of the mantle are largely those of olivine. Additionally, due to the varying temperatures and pressures between the lower and upper mantle, a variety of creep processes can occur with dislocation creep dominating in the lower mantle and diffusional creep occasionally dominating in the upper mantle. However, there is a large transition region in creep processes between the upper and lower mantle and even within each section, creep properties can change strongly with location and thus temperature and pressure. In the power law creep regions, the creep equation fitted to data with n = 3–4 is standard.

The strength of olivine not only scales with its melting temperature, but also is very sensitive to water and silica content. The solidus depression by impurities, primarily Ca, Al, and Na, and pressure affects creep behavior and thus contributes to the change in creep mechanisms with location. While creep behavior is generally plotted as homologous temperature versus stress, in the case of the mantle it is often more useful to look at the pressure dependence of stress. Though stress is simple force over area, defining the area is difficult in geology. Equation demonstrates the pressure dependence of stress. Since it is very difficult to simulate the high pressures in the mantle (1MPa at 300–400 km), the low pressure laboratory data is usually extrapolated to high pressures by applying creep concepts from metallurgy.

$$\left(\frac{\partial \ln \sigma}{\partial P}\right)_{T,\dot{\varepsilon}} = \left(\frac{1}{TT_m}\right) \times \left(\frac{\partial \ln \sigma}{\partial (1/T)}\right)_{P,\dot{\varepsilon}} \times \frac{dT_m}{dP}$$

Most of the mantle has homologous temperatures of 0.65–0.75 and experiences strain rates of 10^{-14}-10^{-16} per second. Stresses in mantle are dependent on density, gravity, thermal expansion coefficients, temperature differences driving convection, and distance convection occurs over, all of which give stresses around a fraction of 3-30MPa. Due to the large grain sizes (at low stresses as high as several mm), it is unlikely that Nabarro-Herring (NH) creep truly dominates. Given the large grain sizes, dislocation creep tends to dominate. 14 MPà is the stress below which diffusional creep dominates and above which power law creep dominates at 0.5Tm of olivine. Thus, even for relatively low temperatures, the stress diffusional creep would operate at is too low for realistic conditions. Though the power law creep rate increases with increasing water content due to weakening, reducing activation energy of diffusion and thus increasing the NH creep rate, NH is generally still not large enough to dominate. Nevertheless, diffusional creep can dominate in very cold or deep parts of the upper mantle. Additional deformation in the mantle can be attributed to transformation enhanced ductility. Below 400 km, the olivine undergoes a pressure induced phase transformation into spinel and can cause more deformation due to the increased ductility. Further evidence for the dominance of power law creep comes from preferred lattice orientations as a result of deformation. Under dislocation creep, crystal structures reorient into lower stress orientations. This does not happen under diffusional creep, thus observation of preferred orientations in samples lends credence to the dominance of dislocation creep.

Mantle Convection in other Celestial Bodies

A similar process of slow convection probably occurs (or occurred) in the interiors of other planets (e.g. Venus, Mars) and some satellites (e.g. Europa, Enceladus).

Plume Tectonics

Plume tectonics is a geoscientific theory that finds its roots in the mantle doming concept which was especially popular during the 1930s and initially did not accept major plate movements and continental drifting. It has survived from the 1970s until today in various forms and presentations. It has slowly evolved into a concept that recognises and accepts large scale plate motions such as envisaged by plate tectonics, but placing them in a framework where large mantle plumes are the major driving force of the system. The initial followers of the concept during the first half of the 20th Century are scientists like Beloussov and van Bemmelen, and recently the concept has gained interest especially in Japan, through new compiled work on palaeomagnetism, and is still advocaded by the group of scientists elaboration upon Earth expansion. It is nowadays generally not accepted as the main theory to explain the driving forces of tectonic plate movements, although numerous modulations on the concept have been proposed.

The theory focuses on the movements of mantle plumes under tectonic plates, viewing them as the major driving force of movements of (parts of) the Earth's crust. In its more modern form, conceived in the 1970s, it tries to reconcile in one single geodynamic model the horizontalistic concept of Plate tectonics, and the verticalistic concepts of mantle plumes, by the gravitational movement of plates away from major domes of the Earth's crust. The existence of various supercontinents in Earth history and their break-up has been associated recently with major upwellings of the mantle.

It is classified together with mantle convection as one of the mechanism that are used to explain the movements of tectonic plates. It also shows affinity with the concept of hot spots which is used in modern day plate teconics to generate a framework of specific mantle upwelling points that are relatively stable throughout time and are used to calibrate the plate movements using their location together with paleomagnetic data. Another affinity is the concept of surge tectonics which envisage flows through the mantle as major driving forces of Plate Tectonics.

Plate Tectonics and Volcanism

The relationships between plate tectonics and volcanism are shown on Figure, magma is formed at three main plate-tectonic settings: divergent boundaries (decompression melting), convergent boundaries (flux melting), and mantle plumes (decompression melting).

The plate-tectonic settings of common types of volcanism. Composite volcanoes form at subduction zones, either on ocean-ocean convergent boundaries (left) or ocean-continent convergent boundaries (right). Both shield volcanoes and cinder cones form in areas of continental rifting. Shield volcanoes form above mantle plumes, but can also form at other tectonic settings. Sea-floor volcanism can take place at divergent boundaries, mantle plumes and ocean-ocean-convergent boundaries.

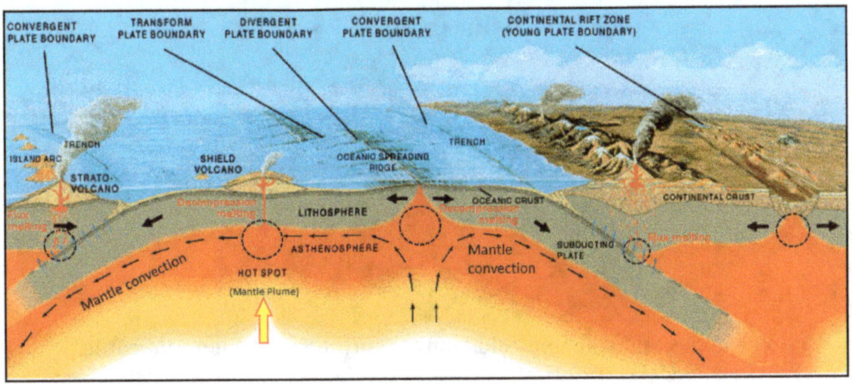

The mantle and crustal processes that take place in areas of volcanism are illustrated in figure At a spreading ridge, hot mantle rock moves slowly upward by convection (cm/year), and within about 60 km of the surface, partial melting starts because of decompression. Over the triangular area shown in figure a, about 10% of the ultramafic mantle rock melts, producing mafic magma that moves upward toward the axis of spreading (where the two plates are moving away from each other). The magma fills vertical fractures produced by the spreading and spills out onto the sea floor to form basaltic pillows (more on that later) and lava flows. There is spreading-ridge volcanism taking place about 200 km offshore from the west coast of Vancouver Island.

At an ocean-continent or ocean-ocean convergent boundary, oceanic crust is pushed far down into the mantle. It is heated up, and while there isn't enough heat to melt the subducting crust, there is enough to force the water out of some of its minerals. This water rises into the overlying mantle where it contributes to flux melting of the mantle rock. The mafic magma produced rises through the mantle to the base of the crust. There it contributes to partial melting of crustal rock, and thus it assimilates much more felsic material. That magma, now intermediate in composition, continues to rise and assimilate crustal material; in the upper part of the crust, it accumulates into plutons. From time to time, the magma from the plutons rises toward surface, leading to volcanic eruptions. Mt. Garibaldi is an example of subduction-related volcanism.

A mantle plume is an ascending column of hot rock (not magma) that originates deep in the mantle, possibly just above the core-mantle boundary. Mantle plumes are thought to rise at approximately 10 times the rate of mantle convection. The ascending column may be on the order of kilometres to tens of kilometres across, but near the surface it spreads out to create a mushroom-style head that is several tens to over 100 kilometres across. Near the base of the lithosphere (the rigid part of the mantle), the mantle plume (and possibly some of the surrounding mantle material) partially melts to form mafic magma that rises to feed volcanoes. Since most mantle plumes are beneath the oceans, the early stages of volcanism typically take place on the sea floor. Over time, islands may form like those in Hawaii.

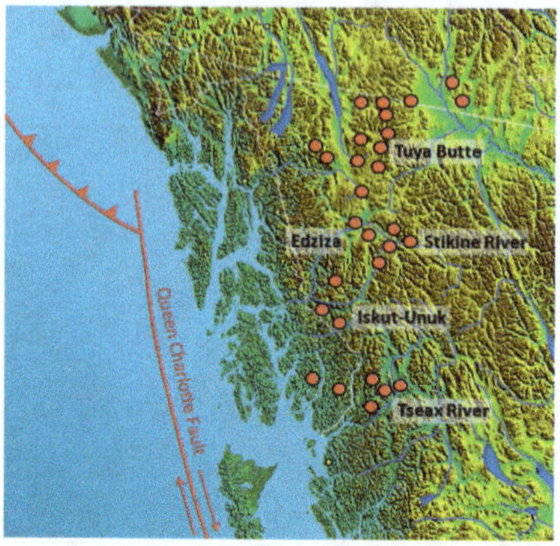

Volcanoes and volcanic fields in the Northern Cordillera Volcanic Province, B.C. Distribution, nature, and origin of Neogene-Quaternary magmatism in the northern Cordilleran volcanic province, Canada.

Volcanic rock at the Tseax River area, northwestern B.C.

Volcanism in northwestern B.C. is related to continental rifting. This area is not at a divergent or convergent boundary, and there is no evidence of an underlying mantle plume. The crust of northwestern B.C. is being stressed by the northward movement of the Pacific Plate against the North America Plate, and the resulting crustal fracturing provides a conduit for the flow of magma from the mantle. This may be an early stage of continental rifting, such as that found in eastern Africa.

References

- Dziewonski, Adam M.; Lekic, Vedran; Romanowicz, Barbara A. (2010). "Mantle Anchor Structure: An argument for bottom up tectonics". Earth and Planetary Science Letters. 299 (1–2): 69–79. Bibcode:2010E&PSL.299...69D. Doi:10.1016/j.epsl.2010.08.013. ISSN 0012-821X
- Plate-tectonics, science: britannica.com, Retrieved 12 April, 2019
- Kobes, Randy and Kunstatter, Gabor."Mantle Convection". Physics Department, University of Winnipeg. Retrieved 3 January 2010
- Plate-tectonics-and-volcanism, chapter, geology: opentextbc.ca, Retrieved 13 May, 2019
- Segev, A (2002). "Flood basalts, continental breakup and the dispersal of Gondwana: evidence for periodic migration of upwelling mantle flows (plumes)". EGU Stephan Mueller Special Publication Series. 2: 171–91. Doi:10.5194/smsps-2-171-2002

6
Impacts of Volcanoes

Volcanoes have a wide range of impacts upon the climate, animal life, plant life and soil. They are also responsible for shaping the surface of the Earth and generating new crust. The topics elaborated in this chapter will help in gaining a better perspective about these diverse impacts of volcanoes as well as their benefits.

Consequences of Volcanic Eruptions

Volcanic eruptions produce spectacular fire fountains, incandescent lava flows, clouds of ash, or bubbling mud pools. They shape the Earth's surface and generate new crust. However, for the people in the immediate vicinity as well as in the wider surroundings, for the climate and for air traffic, these energetic Earth-forming processes are accom-panied by danger and destruction.

Local and Regional Impacts

The immediate surroundings of a volcano experience the most direct effects of an eruption. Explosively ejected ash disperses in the air. Together with volcanic gases it impairs people's ability to breathe. When the ash falls down, it blankets the land, including agriculturally used fields, houses, roads, industrial plants. Also lava flows may seal the area with a layer of rock. In the case of major eruptions, tephra can spread across hundreds of square kilometers and may accumulate to several metres of thickness. Pyroclastic surges (laterally flowing, turbulent density currents consisting of ash, rocks, and gas) can wipe out houses and trees. If loose rock on a volcanic edifice comes into contact with water, lahars can form that devastate the slopes down which they sweep. If a pyroclastic flow, a surge or a lahar reaches a lake or the sea, they may trigger a tsunami – a sudden rise in the water levels of the lake or sea, which leads to abrupt and usually destructive flooding of the shoreline.

Large-scale to Global Impacts

The explosive eruption of Eyjafjallajökull in Iceland in 2010 demonstrated that volcanic ash at higher levels in the atmosphere poses a significant threat to air traffic.

Hot ash particles can clog the turbines of the aircraft and cause engine failure. Huge financial losses for the airlines due to several days of air traffic re-routing or standstills are usually the result. However, the Eyjafjallajökull eruption also gave rise for a range of investigations on the dispersal of the ash cloud using various techniques. It provided an opportunity to improve our understanding on how ash and gas particles travel in the atmosphere. Thereby, it facilitated a better definition of threshold values at which air traffic will be at risk, and what levels of particles can be considered acceptable.

Large eruptions are not without consequences in the long run either. The emission of volcanic gases modifies the composition of the atmosphere. If the gases reach higher altitudes, the effects are particularly strong and long-lasting. While the gas particles in the troposphere, the lowest layer of the atmosphere and approximately 10 km thick, are relatively quickly washed out by rain, they can remain for up to three years in the overlying stratosphere.

Different gas species have different effects in this layer. If sulphur is injected, the stratosphere heats up. The radiation entering from the outside is reflected back into space, so that less radiation arrives on the Earth's surface. The Earth cools down. This mechanism is impressively exemplified by two Indonesian eruptions in the 19th century. As a result of the Tambora eruption in 1815, there was no summer in the following year. In 1883, Krakatoa produced an aerosol cloud which went around the globe and resulted in a "volcanic winter" that lasted for three to four years. This not only caused famines but also visually spectacular, very intensely coloured sunsets, which have been captured in artists' paintings. The most recent example was the eruption of Mount Pinatubo in the Philippines in 1991, followed shortly afterwards by the eruption of Mount Hudson in southern Chile. In the wake of the strong sulphur emissions of these eruptions, global temperatures decreased for two to three years; the polar bear population in the north-east Canadian Hudson Bay exploded.

In contrast, the input of CO2 and halogens such as chlorine and bromine into the stratosphere causes a natural greenhouse effect. These gases destroy ozone at high altitudes, thus amplifying the depletion of the ozone layer. The consequences of stratospheric ozone destruction are well known: more radiation can penetrate the atmosphere; and the Earth warms up.

An Explosive Eruption

If there were an explosive eruption in area the effects of volcanoes could be devastating. During an explosive eruption the volcano will spew lava, magma and volcanic material which could travel several miles away from the mountain. The explosiveness of the eruption could also cause pyroclastic flows which would destroy anything within their path. The ash could also cause severe damage to structures depending on the amount of it and those who breathe it in would be at risk of choking on it.

Problems to Aircraft

Aircraft are some of the things that will feel the effects of volcanoes more than others. That is because the ash spewed by a volcano can negatively affect the engines in an aircraft. To make matters worse the problem has not gotten better. Pilots who have flown through a volcanic eruption complain about the fact that they lose power and when they apply the throttle the engine will become even hotter. When that happens more ash is melted and that creates abrasive material which ends up in the engine.

Effects on Cities and Towns

The effects of volcanoes on cities and towns after an eruption could vary from no effects at all to a catastrophe of immeasurable magnitude. A lot of the effects that a city will feel depend on the size of the explosion or eruption and the atmospheric conditions. It has happened before that the conditions will take away the ash from the city. On the other hand, the same conditions could direct all the harmful effects of the lava, the ashes and volcanic material straight into the city which could cause thousands of deaths or injuries.

To look at the effects of volcanoes in towns and cities it is a good idea to look at recent eruptions. When Mauna Loa erupted in 1950 the lava headed towards the city of Hilo. Another eruption that took place in 1973 took place on Heimaey, Iceland and the entire town was buried under cinder and lava. In 1960 Kilauea erupted and it buried the town of Kapoho and in 1980 when Mount St. Helens erupted the ash fell on several towns in the states of Oregon and Washington. There are several examples of this type of volcano impacts on cities and towns.

Effects on the Environment

There are several ways in which effects of volcanoes can be felt on the environment. Even smaller eruptions can have a measureable effect. That is because while an eruption is taking place gases are released by the volcano. Some of the gases which are released into the air include carbon monoxide, carbon dioxide, sulfur dioxide, fluorine, chlorine, hydrogen sulfide and others. The effects of volcanoes on the environment depend on climate patterns, the overall scale of the eruption and how much the gases spread. Another problem the environment is the ash that is released when a volcano erupts. Depending on the size of the eruption you could be looking at a volcanic winter.

Effects of Supervolcanoes

Out of all the effects of volcanoes on the daily life of people, the one that is the scariest is what is known as the supervolcano. These volcanoes are the ones that will reach the VEI8 and VEI7 rankings in the scale. These are also volcanoes which are very massive and could destroy most of the life on earth if they were to erupt today.

In the case of the supervolcano in Yellowstone Park several states and their citizens would stand no chance against the lava flow and the ash. However, those are not the only people that would be affected. During the explosion a lot of ash would be released into the atmosphere, dropping temperatures by an average of well over 20 degrees Fahrenheit worldwide. That would cause crops and vegetation to die which would mean no food for animals or people.

Benefits of Volcanoes

Volcanoes are renowned for their destructive power. In fact, there are few forces of nature that rival their sheer, awesome might, or have left as big of impact on the human psyche. There's also the Minoan Eruption, the eruption that took place in the 2nd millennium BCE on the isle of Santorini and devastated the Minoan settlement there.

In Japan, Hawaii, South American and all across the Pacific, there are countless instances of eruptions taking a terrible toll. And who can forget modern-day eruptions like Mount St. Helens? But would it surprise you to know that despite their destructive power, volcanoes actually come with their share of benefits? From enriching the soil to creating new landmasses, volcanoes are actually a productive force as well.

Volcanic Land Formations

In addition to scattering ash over large areas of land, volcanoes also push material to the surface that can result in the formation of new islands. For example, the entire Hawaiian chain of islands was created by the constant eruptions of a single volcanic hot spot. Over hundreds of thousands of years, these volcanoes breached the surface of the ocean becoming habitable islands, and rest stops during long sea journeys.

The island of Santorini, Greece.

This is the case all across the Pacific, were island chains such as Micronesia, the Ryukyu Islands (between Taiwan and Japan), the Aleutian Islands (off the coast of Alaska), the

Mariana Islands, and Bismark Archipelago were all formed along arcs that are parallel and close to a boundary between two converging tectonic plates.

Much the same is true of the Mediterranean. Along the Hellenic Arc (in the eastern Mediterranean), volcanic eruptions led to the creation of the Ionian Islands, Cyprus and Crete. The nearby South Aegean Arc meanwhile led to the formation of Aegina, Methana, Milos, Santorini and Kolumbo, and Kos, Nisyros and Yali. And in the Caribbean, volcanic activity led to the creation of the Antilles archipelago.

Where these islands formed, unique species of plants and animals evolved into new forms on these islands, creating balanced ecosystems and leading to new levels of biodiversity.

Volcanic Minerals and Stones

Another benefits to volcanoes are the precious gems, minerals and building materials that eruptions make available. For instance, stones like pumice volcanic ash and perlite (volcanic glass) are all mined for various commercial uses. These include acting as abrasives in soaps and household cleaners. Volcanic ash and pumice are also used as a light-weight aggregate for making cement.

The finest grades of these volcanic rocks are used in metal polishes and for woodworking. Crushed and ground pumice are also used for loose-fill insulation, filter aids, poultry litter, soil conditioner, sweeping compound, insecticide carrier, and blacktop highway dressing.

The roof of the Pantheon, as seen from nearby rooftops in Roe.

Perlite is also used as an aggregate in plaster, since it expands rapidly when heated. In precast walls, it too is used as an aggregate in concrete. Crushed basalt and diasbase are also used for road metal, railroad ballast, roofing granules, or as protective arrangements for shorelines (riprap). High-density basalt and diabase aggregate are used in the concrete shields of nuclear reactors.

Hardened volcanic ash (called tuff) makes an especially strong, lightweight building material. The ancient Romans combined tuff and lime to make a strong, lightweight concrete for walls, and buildings. The roof of the Pantheon in Rome is made of this very type of concrete because it's so lightweight.

Precious metals that are often found in volcanoes include sulfur, zinc, silver, copper, gold, and uranium. These metals have a wide range of uses in modern economies, ranging from fine metalwork, machinery and electronics to nuclear power, research and medicine. Precious stones and minerals that are found in volcanoes include opals, obsidian, fire agate, flourite, gypsum, onyx, hematite, and others.

Global Cooling

Volcanoes also play a vital role in periodically cooling off the planet. When volcanic ash and compounds like sulfur dioxide are released into the atmosphere, it can reflect some of the Sun's rays back into space, thereby reducing the amount of heat energy absorbed by the atmosphere. This process, known as "global dimming", therefore has a cooling effect on the planet.

Sarychev volcano, (located in Russia's Kuril Islands, northeast of Japan) in an early stage of eruption.

The link between volcanic eruptions and global cooling has been the subject of scientific study for decades. In that time, several dips have been observed in global temperatures after large eruptions. And though most ash clouds dissipate quickly, the occasional prolonged period of cooler temperatures have been traced to particularly large eruptions.

Because of this well-established link, some scientists have recommended that sulfur dioxide and other be released into the atmosphere in order to combat global warming, a process which is known as ecological engineering.

Hot Springs and Geothermal Energy

Another benefit of volcanism comes in the form of geothermal fields, which is an area of the Earth characterized by a relatively high heat flow. These fields, which are the result of present, or fairly recent magmatic activity, come in two forms. Low temperature fields (20-100°C) are due to hot rock below active faults, while high temperature fields (above 100°C) are associated with active volcanism.

Geothermal fields often create hot springs, geysers and boiling mud pools, which are often a popular destination for tourists. But they can also be harnessed for geothermal energy, a form of carbon-neutral power where pipes are placed in the Earth and channel steam upwards to turn turbines and generate electricity.

Steam rising from the Nesjavellir Geothermal Power Station in Iceland.

In countries like Kenya, Iceland, New Zealand, the Phillipines, Costa Rica and El Salvador, geothermal power is responsible for providing a significant portion of the country's power supply – ranging from 14% in Costa Rica to 51% in Kenya. In all cases, this is due to the countries being in and around active volcanic regions that allow for the presence of abundant geothermal fields.

Outgassing and Atmospheric Formation

But by far, the most beneficial aspect of volcanoes is the role they play in the formation of a planet's atmosphere. In short, Earth's atmosphere began to form after its formation 4.6 billion eyars ago, when volcanic outgassing led to the creation of gases stored in the Earth's interior to collect around the surface of the planet. Initially, this atmosphere consisted of hydrogen sulfide, methane, and 10 to 200 times as much carbon dioxide as today's atmosphere.

After about half a billion years, Earth's surface cooled and solidified enough for water to collect on it. At this point, the atmosphere shifted to one composed of water vapor, carbon dioxide and ammonia (NH^3). Much of the carbon dioxide dissolved into the oceans, where cyanobacteria developed to consume it and release oxygen as a byproduct. Meanwhile, the ammonia began to be broken down by photolysis, releasing the hydrogen into space and leaving the nitrogen behind.

Roughly 2.5 billion years ago, towards the end of the Archaean Era, oxidation of our atmosphere began.

Another key role played by volcanism occurred 2.5 billion years ago, during the boundary between the Archaean and Proterozoic Eras. It was at this point that oxygen began to appear in our oxygen due to photosynthesis – which is referred to asthe "Great Oxidation Event". However, according to recent geological studies, biomarkers indicate that oxygen-producing cyanobacteria were releasing oxygen at the same levels there are today. In short, the oxygen being produced had to be going somewhere for it not to appear in the atmosphere.

The lack of terrestrial volcanoes is believed to be responsible. During the Archaean Era, there were only submarine volcanoes, which had the effect of scrubbing oxygen from the atmosphere, binding it into oxygen containing minerals. By the Archaean/Proterozoic boundary, stabilized continental land masses arose, leading to terrestrial volcanoes. From this point onward, markers show that oxygen began appearing in the atmosphere.

Volcanism also plays a vital role in the atmospheres of other planets. Mercury's thin exosphere of hydrogen, helium, oxygen, sodium, calcium, potassium and water vapor is due in part of volcanism, which periodically replenishes it. Venus' incredibly dense atmosphere is also believed to be periodically replenished by volcanoes on its surface.

And Io, Jupiter's volcanically active moon, has an extremely tenuous atmosphere of sulfur dioxide (SO^2), sulfur monoxide (SO), sodium chloride (NaCl), sulfur monoxide (SO), atomic sulfur (S) and oxygen (O). All of these gases are provided and replenished by the many hundreds of volcanoes situated across the moon's surface.

Effects of Volcanoes on Climate

Volcanic eruptions are responsible for releasing molten rock, or lava, from deep within the Earth, forming new rock on the Earth's surface. But eruptions also impact the atmosphere.

A huge cloud of volcanic ash and gas rises above Mount Pinatubo, Philippines. Three days later, the volcano exploded in the second-largest volcanic eruption on Earth in the 20th century.

The gases and dust particles thrown into the atmosphere during volcanic eruptions have influences on climate. Most of the particles spewed from volcanoes cool the planet by shading incoming solar radiation. The cooling effect can last for months to years depending on the characteristics of the eruption. Volcanoes have also caused global warming over millions of years during times in Earth's history when extreme amounts of volcanism occurred, releasing greenhouse gases into the atmosphere.

Even though volcanoes are in specific places on Earth, their effects can be more widely distributed as gases, dust, and ash get into the atmosphere. Because of atmospheric circulation patterns, eruptions in the tropics can have an effect on the climate in both hemispheres while eruptions at mid or high latitudes only have impact the hemisphere they are within.

Particles of Dust and Ash

Volcanic ash or dust released into the atmosphere during an eruption shade sunlight and cause temporary cooling. Larger particles of ash have little effect because they fall out of the air quickly. Small ash particles form a dark cloud in the troposphere that shades and cools the area directly below. Most of these particles fall out of the atmosphere within rain a few hours or days after an eruption. But the smallest particles of dust get into the stratosphere and are able to travel vast distances, often worldwide. These tiny particles are so light that they can stay in the stratosphere for months, blocking sunlight and causing cooling over large areas of the Earth.

Sulfur

Often, erupting volcanoes emit sulfur dioxide into the atmosphere. Sulfur dioxide is much more effective than ash particles at cooling the climate. The sulfur dioxide moves into the stratosphere and combines with water to form sulfuric acid aerosols. The sulfuric acid makes a haze of tiny droplets in the stratosphere that reflects incoming solar radiation, causing cooling of the Earth's surface. The aerosols can stay in the stratosphere for up to three years, moved around by winds and causing significant cooling worldwide. Eventually, the droplets grow large enough to fall to Earth.

Greenhouse Gases

Volcanoes also release large amounts of greenhouse gases such as water vapor and carbon dioxide. The amounts put into the atmosphere from a large eruption doesn't change the global amounts of these gases very much. However, there have been times during Earth history when intense volcanism has significantly increased the amount of carbon dioxide in the atmosphere and caused global warming.

Effects on Plants and Animals

Plants are destroyed over a wide area, during an eruption. The good thing is that volcanic soil is very rich, so once everything cools off, plants can make a big comeback.

Livestock and other mammals have been killed by lava flows, pyroclastic flows, tephra falls, atmospheric effects, gases, and tsunami. They can also die from famine, forest fires, and earthquakes caused by or related to eruptions.

Mount St. Helens provides an example. The Washington Department of Game estimated that 11,000 hares, 6,000 deer, 5,200 elk, 1,400 coyotes, 300 bobcats, 200 black bears, and 15 mountain lions died from the pyroclastic flows of the 1980 eruption.

Aquatic life can be affected by an increase in acidity, increased turbidity, change in temperature, and/or change in food supply. These factors can damage or kill fish. Eruptions can influence bird migration, roosting, flying ability, and feeding activity. The impact of eruptions on insects depends on the size of the eruption and the stage of growth of the insect. For example, ash can be very abrasive to wings.

How quickly do plants begin to grow back? it depends on how much rain falls in the particular area. For example, on the rainy side of the island of Hawaii, flows that are only 2 years old already have ferns and small trees growing on them. Probably in 10 years they'll be covered by a low forest. On the dry side of Hawaii there are flows a couple hundred years old with hardly a tuft of grass in sight. This means that when you are looking at old lava flows and trying to determine how old they are based on the amount of vegetation, you have to take the climate into effect as well.

Long Term Effects

The long-term effects of an eruption on wildlife are usually quite small. Certainly at Mt. St. Helens scientists saw that both plants and animals returned to the utterly devastated areas within only a year or so of the eruption.It is usually the short-term effects that are really bad. For example, there was a very big eruption of Santa Maria volcano (Guatemala) in 1902. The eruption itself killed a few hundred to perhaps 1500 people as well as thousands of birds. Pretty soon there were so many insects including disease-carrying mosquitoes that eventually 3000-6000 people died from malaria.

Extinction of Dinosaurs

There are various variations on the main theory. In general it is proposed that volcanic activity put so much ash and/or gas into the atmosphere that the earth's temperature either got too hot for the dinosaurs or got too cold for the dinosaurs. It sounds kind of funny that either can happen but it is true. If the ash particles are really small (<2 microns) then they block out incoming sunlight and the earth gets cool. If they are bigger than 2 microns (but still pretty small) then they let sunlight in but don't let heat radiation from the surface out, and the earth gets warm. Anyway, if you have enough large explosive eruptions, then the theory says that there will be enough ash in the stratosphere to have one of these effects. You need an eruption (or series of eruptions) that is much bigger than anything we have ever witnessed. The reason that you need to put the ash into the stratosphere is that if it is only in the troposphere (where weather clouds are), then it will get rained out very quickly and it won't be around long enough to have a climatic effect.

Of course the more famous idea is that a huge meteorite came in and hit the earth, throwing up enough gas and dust into the stratosphere to have the same heating or cooling effect. One line of support for this is that at the geologic time boundary where the dinosaurs died out (the Cretaceous-Tertiary boundary) there is a layer of clay that is rich in an element called iridium. Iridium is not very common on Earth, but it is proposed to be more abundant in asteroids and meteorites. One way to produce such a layer at the same instant that the dinosaurs died out is therefore to have a meteorite bring it in.

One major problem with the volcanic hypothesis is that volcanoes, especially the explosive ones, don't produce much iridium. Basaltic volcanoes, such as those here in Hawaii produce more iridium but they are not very explosive.

A more recent idea that tries to get around these problems is that instead of a huge explosive eruption, you have a long-term basaltic eruption that mainly puts SO_2 gas into the troposphere. The gas will be converted into small droplets of sulfuric acid which will block incoming sunlight. Because it is only in the troposphere much of the acid may get rained out, but if you have an eruption that continues long enough it can keep up with the rain to produce an Earth-covering haze.

There are places on Earth where huge volumes of basaltic lavas are found. They are called flood basalts, and the most famous are the Columbia River Basalts in Washington/Oregon, and the Deccan Traps in India. The name "flood basalts" gives an indication of how most people consider them to be erupted, namely as huge fast-moving floods of basalt. However, recent work by a number of scientists here at the University of Hawaii (including Steve Self, George Walker, Thorvaldur Thordarson, and Sarah Finnemore) have shown that these flood basalts look more like the slow-moving type of basalt lava (pahoehoe) than the fast-moving type ('a'a). This leads next to the conclusion that perhaps these flood basalts were not emplaced as huge floods in short periods of time but rather as slower-moving flows over a long period of time (such as 1-2 hundred years). The eruptions would still have been much bigger than those we see here in Hawaii, however.

Flood Basalt

A flood basalt is the result of a giant volcanic eruption or series of eruptions that covers large stretches of land or the ocean floor with basalt lava. Flood basalt provinces such as the Deccan Traps of India are often called traps, after the Swedish word trappa, due to the characteristic stairstep geomorphology of many associated landscapes. Michael R. Rampino and Richard Stothers (1988) cited eleven distinct flood basalt episodes occurring in the past 250 million years, creating large volcanic provinces, plateaus, and mountain ranges. However, more have been recognized such as the large Ontong Java Plateau, and the Chilcotin Group, though the latter may be linked to the Columbia River Basalt Group. Large igneous provinces have been connected to five mass extinction events, and may be associated with bolide impacts.

Formation of Flood Basalt

Moses Coulee in the US showing multiple flood basalt flows of the Columbia River Basalt Group. The upper basalt is Roza Member, while the lower canyon exposes Frenchmen Springs Member basalt.

The formation and effects of a flood basalt depend on a range of factors, such as continental configuration, latitude, volume, rate, duration of eruption, style and setting (continental vs. oceanic), the preexisting climate, and the biota resilience to change.

One proposed explanation for flood basalts is that they are caused by the combination of continental rifting and its associated decompression melting, in conjunction with a mantle plume also undergoing decompression melting, producing vast quantities of a tholeiitic basaltic magma. These have a very low viscosity, which is why they 'flood' rather than form taller volcanoes. Another explanation is that they result from the release, over a short period, of melt that has accumulated in the mantle over a long period.

Lava flows at Holuhraun, Iceland, September, 2014.

The Deccan Traps of central India, the Siberian Traps, and the Columbia River Plateau of western North America are three regions covered by prehistoric flood basalts. The Mesoproterozoic Mackenzie Large Igneous Province in Canada contains the Coppermine River flood basalts related to the Muskox layered intrusion. The maria on the Moon are additional, even more extensive, flood basalts. Flood basalts on the ocean floor produce oceanic plateaus.

The surface covered by one eruption can vary from around 200,000 km^2 (Karoo) to 1,500,000 km^2 (Siberian Traps). The thickness can vary from 2000 metres (Deccan Traps) to 12,000 m (Lake Superior). These are smaller than the original volumes due to erosion.

Petrography

Ethiopian Highlands basalt.

Flood basalts have tholeiite and olivine compositions (according to the classification of Yoder and Tilley). The composition of the basalts from the Paraná is fairly typical of that of flood basalts; it contains phenocrysts occupying around 25% of the volume of rock in a fine-grained matrix. These phenocrysts are pyroxenes (augite and pigeonite), plagioclases, opaque crystals such as titanium rich magnetite or ilmenite, and occasionally some olivine. Sometimes more differentiated volcanic products such as andesites, dacites and rhyodacites have been observed, but only in small quantities at the top of former magma chambers.

Structures

Subaerial flood basalts can be of two kinds:

- With a smooth or twisted surface (Pāhoehoe): very compact surface; vesicles (gas bubbles) are rare. Degassing was easy (magma maintained at a high temperature and more fluid in a chamber of a size such that confining pressures did not confine gases to the melt before expulsion). Such lava flows may form underground rivers; when degassing fractures and conduits are present, very large flows may reach the surface.
- With a chaotic surface ('A'ā): the basalt flood is very rich in bubbles of gas, with an irregular, fragmental surface. Degassing was difficult (less fluid magma expelled from a rift with no chance of progressive expansion in a hot chamber; the degassing took place closer to the surface where the flow forms a crust which cracks under the pressure of the gases in the flow itself and during more rapid cooling).

In the Massif Central in Auvergne, France, there is a good example of chaotic lava flow, produced by eruptions from Puy de la Vache and Puy de Lassolas.

Large Igneous Provinces

Large Igneous Provinces (LIPs) were originally defined as including voluminous outpourings, predominantly of basalt, over geologically very short durations. This definition did not specify minimum size, duration, petrogenesis, or setting. A new attempt to refine classification focuses on size and setting. LIPs characteristically cover large areas and the great bulk of the magmatisim occurs in about less than 1 Ma. Principal LIPs in the ocean basins include Oceanic Volcanic Plateaus (OPs) and Volcanic Passive Continental Margins. Oceanic flood basalts are LIPs distinguished from oceanic plateaus by some investigators because they do not form morphologic plateaus, being neither flat-topped nor elevated more than 200 m above the seafloor. Examples include the Caribbean, Nauru, East Mariana, and Pigafetta provinces. Continental flood basalts (CFBs) or plateau basalts are continental manifestations, or traps referencing the step-like geomorphology of eroded flow layers. (e.g. Deccan Traps and Siberian Traps).

Geochemistry

Geochemical analysis of the major oxides reveals a composition close to that of mid-ocean ridge basalts (MORB) but also close to that of ocean island basalts (OIB). These are in fact tholeiites with a silicon dioxide percentage close to 50%.

Multiple flood basalt flows of the Chilcotin Group, British Columbia, Canada

Two kinds of basaltic flood basalts can be distinguished:

- Those poor in P_2O_5 and in TiO_2, called low phosphorus and titanium;
- Those rich in P_2O_5 and in TiO_2, called high phosphorus and titanium.

The isotopic ratios $^{87}Sr/^{86}Sr$ and $^{206}Pb/^{204}Pb$ are different from that observed in general, which shows that the basalt flood magma was contaminated as it passed through the continental crust. It is this contamination that explains the difference between the two kinds of basalt mentioned above. The low phosphorus and titanium type has an excess of elements from the crust such as potassium and strontium.

The content in incompatible elements of flood basalts is lower than that of ocean island basalts, but higher than that of mid-ocean ridge basalts.

Effect of Volcano on Soil

Volcanoes can be destructive, but volcanic action can lead to the formation of fertile soils. Volcanic activity is traditionally associated with the destruction of lives and property because events such as eruption can decimate all living things. However, the process of eruption is characterized by different processes which ultimately lead to the formation of new materials including rocks and soil with volcanic soils formed through this process. They are among the most fertile in the world.

Soil Fertility

Volcanic eruptions result in the expulsion of ash which is spread out across wide areas both near and far from the volcanic mountain. Therefore, depending on the composition of the magma from a particular volcanic mountain, the resulting ash will have distinct amounts of soil nutrients. Silica and oxygen take up the largest percentage of elements released in the event of a volcanic eruption; other elements include carbon dioxide, sulfur dioxide, and hydrogen chloride.

Rock Particles

Another factor that contributes to elevated levels of fertility in volcanic soils is the bits of rock such as feldspar which have an abundance of minerals such as iron, magnesium, and potassium. These are some of the most important minerals found in fertilizers to improve the quality of agricultural soil. Because of the presence of these rock particles being incorporated into the soil, the regions that have volcanic soil, for instance, the valleys and mountain slopes are typically fertile.

Basalt

Geologists and soil scientists have also found that volcanic soils are primarily composed of basalt. Crystals of this particular type of soils are small and break down with ease in addition to having high quantities of iron. Other elements such as phosphorus and calcium in these easily weathered soils significantly contribute to overall fertility of the soils.

Deposition and Weathering

In regions near Mount Vesuvius, there are expansive tracts of fertile lands as a result of the volcanic activity thousands of years ago. The fertility of the soil in this region is attributed to the deposition of volcanic materials rich in minerals which are then broken down or weathered by erosion or action of water. In this state, they are easily absorbed into the soil to become a continuous supply of essential nutrients to the plants found in the region.

Volcanic Activity and Agriculture

Volcanic action leading to the emergence of fertile soils has greatly contributed to agriculture as an economic activity. In Hawaii, for instance, agriculture gained ground between the 15th and 18th century due to volcanicity and has grown ever since. Cultivation of food crops including sweet potatoes led to the growth of chiefdoms which had a huge influence on the societies of Hawaii as seen today.

Impacts of Volcanic Gases on Climate, Environment and People

The molten rock, or magma, that lies beneath volcanoes contains abundant gases that are released to the surface before, during, and after eruptions. These gases range from relatively benign low-temperature steam to thick hot clouds of choking acid fume jetting from the earth.

The concentrations of different volcanic gases can vary considerably from one volcano to the next. Water vapor is typically the most abundant volcanic gas, followed by carbon dioxide and sulfur dioxide. Other principal volcanic gases include hydrogen sulfide, hydrogen chloride, and hydrogen fluoride. A large number of minor and trace gases are also found in volcanic emissions for example, hydrogen, carbon monoxide, and volatile metal chlorides.

A significant component of volcanic gas research involves measuring the quantities of gas that volcanoes release into the atmosphere. Huge amounts of volcanic gas, aerosol droplets, and ash are injected into the stratosphere during major explosive eruptions. Some gases, such as carbon dioxide, are greenhouse gases that promote global warming, while others, like sulfur dioxide, can cause global cooling, ozone destruction, and polluted air known as volcanic smog or Vog'. Studies of volcanic emissions allow scientists to compare volcanic gas output to emissions from man-made sources and to assess the effects of both past and future eruptions on the Earth's climate.

Volcanic gas can also be directly harmful to humans, animals, plants, agricultural crops, and property. Usually, the hazards from volcanic gases are most severe in the areas immediately surrounding volcanoes, especially on volcano flanks downwind of active vents and fumaroles. But these hazards can persist for long distances downwind following large eruptions, or from volcanoes erupting gas-rich magma.

Health hazards can range from minor to life threatening. Exposure to acid gases such as sulfur dioxide, hydrogen sulfide, and hydrogen chloride can damage eyes and mucous membranes along with the respiratory system and, under extreme conditions, can lead to death. The health effects from exposure to prolonged periods of volcanic smog are not well documented, but the common complaints include headaches, fatigue, respiratory difficulties, and allergic reactions. The toxicity of carbon monoxide is well known, although it is rarely abundant enough in volcanic gases to cause serious problems. A very serious hazard can occur under certain conditions from volcanic emissions of carbon dioxide. Carbon dioxide is heavier than air and can collect in low and poorly ventilated places. Nearly two thousand people have died of carbon dioxide asphyxiation near volcanoes in the past two decades, most of them in Cameroon, Africa, and in Indonesia. Recently, there have been cases of near asphyxiation from carbon dioxide emissions at Mammoth Mountain, a young volcano on the eastern front of the Sierra Nevada mountains in central California.

Volcanic gases can also severely damage vegetation. Direct exposure to concentrated volcanic gas or long-term exposure to dilute volcanic gas is lethal to most types of foliage. Since 1990, areas of dying forest around Mammoth Mountain have grown in size and number because of high concentrations of carbon dioxide in the soils and now occupy more than 100 acres.

Fume clouds from volcanoes also contain water droplets in which acid gases have dissolved. Most of these droplets eventually fall to earth as acid rain, but while they remain in the atmosphere, they can also cause corrosion of aircraft surface materials and engines. Utility lines, communications equipment, farm machinery, vehicles, and other metal objects corrode when exposed to volcanic gases or acid rain. Persistent acid rain causes galvanized nails or lead solder in water catchment systems to deteriorate and release toxic metals into drinking water. Hydrogen fluoride gas emitted from a volcano can attach itself to ash particles. When these particles fall to earth, chronic and fatal fluorosis can follow when grazing animals eat contaminated forage.

References

- Consequences-of-volcanic-eruptions, natural-hazards, basic-knowledge, en: eskp.de, Retrieved 14 June, 2019
- David P.G. Bond; Paul B. Wignall (2014). "Large igneous provinces and mass extinctions: An update". GSA Special Papers. 505: 29–55. Doi:10.1130/2014.2505(02). Archived from the original on 2014-08-06
- Effects-volcanoes: basicplanet.com, Retrieved 15 July, 2019
- Winter, John (2010). Principles of Igneous and Metamorphic Petrology (2nd ed.). New York: Prentice Hall. Pp. 301–302. ISBN 9780321592576
- Benefits-of-volcanoes: universetoday.com, Retrieved 16 August, 2019
- Wallace, P. J.; Frey, F. A.; Weis, D.; Coffin, M. F. (2002). "Origin and Evolution of the Kerguelen Plateau, Broken Ridge and Kerguelen Archipelago: Editorial". Journal of Petrology. 43 (7): 1105–1108. Bibcode:2002jpet...43.1105W. Doi:10.1093/petrology/43.7.1105
- How-volcanoes-influence-climate, shortcontent: scied.ucar.edu, Retrieved 17 January, 2019
- How-do-volcanoes-affect-plants-and-animals: volcano.oregonstate.edu, Retrieved 18 February, 2019

Permissions

All chapters in this book are published with permission under the Creative Commons Attribution Share Alike License or equivalent. Every chapter published in this book has been scrutinized by our experts. Their significance has been extensively debated. The topics covered herein carry significant information for a comprehensive understanding. They may even be implemented as practical applications or may be referred to as a beginning point for further studies.

We would like to thank the editorial team for lending their expertise to make the book truly unique. They have played a crucial role in the development of this book. Without their invaluable contributions this book wouldn't have been possible. They have made vital efforts to compile up to date information on the varied aspects of this subject to make this book a valuable addition to the collection of many professionals and students.

This book was conceptualized with the vision of imparting up-to-date and integrated information in this field. To ensure the same, a matchless editorial board was set up. Every individual on the board went through rigorous rounds of assessment to prove their worth. After which they invested a large part of their time researching and compiling the most relevant data for our readers.

The editorial board has been involved in producing this book since its inception. They have spent rigorous hours researching and exploring the diverse topics which have resulted in the successful publishing of this book. They have passed on their knowledge of decades through this book. To expedite this challenging task, the publisher supported the team at every step. A small team of assistant editors was also appointed to further simplify the editing procedure and attain best results for the readers.

Apart from the editorial board, the designing team has also invested a significant amount of their time in understanding the subject and creating the most relevant covers. They scrutinized every image to scout for the most suitable representation of the subject and create an appropriate cover for the book.

The publishing team has been an ardent support to the editorial, designing and production team. Their endless efforts to recruit the best for this project, has resulted in the accomplishment of this book. They are a veteran in the field of academics and their pool of knowledge is as vast as their experience in printing. Their expertise and guidance has proved useful at every step. Their uncompromising quality standards have made this book an exceptional effort. Their encouragement from time to time has been an inspiration for everyone.

The publisher and the editorial board hope that this book will prove to be a valuable piece of knowledge for students, practitioners and scholars across the globe.

Index

www.ingramcontent.com/pod-product-compliance
Lightning Source LLC
LaVergne TN
LVHW080609210926
848485LV00003B/128

* 9 7 8 1 6 4 1 1 6 5 7 2 3 *